Essentials of Ultramicrotomy

Anjali Haloi

Essentials of Ultramicrotomy

Springer

Anjali Haloi
Sophisticated Analytical Instrument Facility
North-Eastern Hill University
Shillong, Meghalaya, India

ISBN 978-981-97-4793-1 ISBN 978-981-97-4791-7 (eBook)
https://doi.org/10.1007/978-981-97-4791-7

This Springer imprint is published by the registered company Springer Nature Singapore Pte Ltd.
The registered company address is: 152 Beach Road, #21-01/04 Gateway East, Singapore 189721, Singapore

From the Author's Desk

I firmly believe that learning is an ongoing and infinite journey. The valuable experiences I acquired through the daily operations and maintenance of Ultramicrotomy equipment are meticulously detailed in this book. This book presents the working experience of more than 16 years in the Transmission Electron Microscopy (TEM) division of the Sophisticated Analytical Instrument Facility (SAIF), North-Eastern Hill University (NEHU), India, where my primary duty involves handling ultramicrotome for TEM section other than biological and material sample processing, staining, and viewing of sample in TEM. The duty also includes training participants, demonstrating to the study tour students from different colleges and universities, demonstrating practicals during regular workshops, and presenting in seminars and webinars. Through these actions, I have gathered valuable occurrences by encountering the partaker's queries and doubts and solving them.

The motivation to write this book arose from the observation that our facility regularly receives a multitude of biological samples for TEM from across the country. It became apparent that many of these institutions possess TEM facilities either on their premises or in nearby locations. According to many users who have utilised our facility, one of the main reasons out of others is that those facilities cannot process biological samples, especially the ultrathin sectioning or ultramicrotomy part, prior to TEM viewing, as it involves many tedious steps. Understanding this issue based on my initial stages of learning ultramicrotomy, the maximum resolution of a microscope. Abbe's equation may be discussed broadly in Chap. 2 of this book.

- In 1900, it was found that a light microscope has a resolution limit of 0.2 μm (2000 angstroms) and can magnify an object up to 500–1000×. Four years after overcoming this limitation, Carl Zeiss introduced the first commercial UV microscope, which had a resolution twice as powerful as a visible light microscope.
- In 1931, two German scientists, Ernst Ruska, a physicist, and Max Knoll, an electrical engineer from the University of Berlin, discovered a way to achieve a resolution more significant than light. This prototype produced a magnification of 400 power and was the first device to show what was possible with electron microscopy. They found that transmitted electrons through a specimen can form an image instead of light. This discovery led to the first transmission electron microscope (TEM), where the electrons are pointed directly at the sample and pass through the object to create the image. Modern TEMs can visualise objects as small as the diameter of an atom.
- In 1932, Frits Zernike, a physicist at the University of Groningen, Holland, discovered the phase principle and described its use in microscopy in 1935. In 1953, he won the Nobel Prize in Physics for this work. In this technique, he used a disk in the condenser to separate direct light from the specimen from diffracted light. He increased the phase difference between the direct and diffracted light by using a special plate in the back focal plane of the objective lens. As a result, interference between the direct and diffracted light increased, and the intermediate image plane produced amplitude contrast that the microscopist could see. By using interference rather than absorption of light, transparent samples, such as cells, can be imaged without staining techniques.
- In 1935, the German scientist Max Knoll built the first "scanning microscope", and Ernst Ruska 1942 also contributed to the building of the scanning electron microscope (SEM), which transmits a beam of electrons across the surface of a

specimen. However, there are many other contributors to the invention of the SEM principle. SEM can image topography and obtain material information on conductive specimens using a focused beam of high-energy electrons. The electron beam is deflected in a magnetic field and performs a scanning movement in a raster pattern to capture the specimens' surface. For imaging purposes, interaction phenomena of the electron beam with the specimen, like emission of secondary electrons (SE) or backscattered electrons (BSE), are detected and converted to grey values. A frequency analysis of X-rays reveals information about the present material. SEM images have a three-dimensional appearance, which is very useful when examining the surface structure of a sample.

– In 1957, Marvin Minsky patents the principle of confocal imaging, where he developed the principle of use of illumination—and detection-side pinhole apertures in the same conjugate image plane, making them "confocal". In that configuration, one pinhole was placed in front of a zirconium arc light source to provide a point of light focused on the sample by an objective lens. The second objective lens focused the illuminated sample point onto the second pinhole in front of the detector. This "double focusing" system rejects the out-of-focus rays from the illuminated sample, so they do not reach the detector, which was a low-noise photomultiplier. To build the resulting image, the stage could be moved in *x* and *y* to scan the sample through the illumination point. The primary functions of a confocal microscope are to produce a point source of light and reject out-of-focus light, which provides the ability to image deep into high-resolution tissues and optical sectioning for 3D reconstructions of imaged samples.
– In 1962, Osamu Shimomura, Frank Johnson, and Yo Saiga discovered green fluorescent protein (GFP) while purifying the bioluminescent protein aequorin from jellyfish, Aequorea victoria. GFP fluoresces bright green when exposed to blue light. Morise and colleagues in 1974 provided insight into the

fluorescent properties of GFP, which was shown to emit green light on energy transfer from aequorin.

- In 1972, Godfrey Hounsfield and Allan Cormack developed the computerised axial tomography (CAT) or computerised tomography (CT) scanner. However, the mathematical theory of Johann Radon in 1917, called the "Radon transform", brought the technology to life. Another mathematical advancement that Hounsfield built on is the "Algebraic Reconstruction Technique", formulated by Polish mathematician Stefan Kaczmarz in 1937. Hounsfield adopted both theories to create one of the most significant advancements in medical history. With the help of a computer, the device combines many X-ray images to generate cross-sectional views and three-dimensional images of internal organs and structures. Godfrey Hounsfield co-invented the technology with physicist Dr. Allan Cormack. Both researchers were later jointly awarded the 1979 Nobel Prize in Physiology and Medicine. By 1981, Hounsfield was knighted and became Sir Godfrey Hounsfield. He contributed enormously towards the diagnosis of neurological and other disorders by his invention of this instrument.
- In 1978, Thomas and Christoph Cremer developed a Confocal laser scanning microscope, which scans an object using a focused laser beam. *Simultaneously, Colin Sheppard also developed the technique with a theory of image formation. Tony Wilson, Brad Amos, and John White nurtured the concept during the late 1980s and later demonstrated the utility of confocal imaging in examining fluorescent biological specimens.* Confocal microscopes are scanning microscopes, the primary feature of which is a substantially improved longitudinal (axial) resolution by suppressing image contributions from light outside a relatively thin plane. This allows one to acquire three-dimensional images, e.g., of transparent biological and medical samples—effectively applying a kind of "optical sectioning" instead of physical sectioning. With such microscopes, one can even observe tiny details within living cells.

Therefore, such instruments have become necessary for different research and industrial inspection areas.

- In 1981, scanning tunnelling microscope (STM) was invented by Gerd Binnig and Heinrich Rohrer. The STM "sees" by measuring interactions between atoms rather than using light or electrons. Binnig and Rohrer's invention enabled scientists to visualise molecules and their atoms within materials. This instrument opened the door to nanotechnology, electrochemistry, semiconductor science, and molecular biology research. The microscope has improved the level of precision in mechanical design and, as a result, has achieved more explicit images. This advanced microscope gives access to the nanoscale world of individual atoms and molecules for the first time. Because of its high-resolution imaging power and broad applicability, this microscope has found significant applications in physics, chemistry, engineering, and materials science. For an STM, a good resolution is considered 0.1 nm of lateral and 0.01 nm of depth resolution. Gerd Binnig and Heinrich Rohrer were awarded the Nobel Prize in Physics for this invention in 1986.
- In 1986, Ernst Ruska (for his work on the electron microscope) and Gerd Binnig and Heinrich Rohrer (for the scanning tunnelling microscope) were jointly awarded the Nobel Prize in Physics for their ingenious idea in developing the world of microscopes.
- 1992—green fluorescent protein (GFP) cloning reports by Douglas Prasher. This opens the way to the widespread use of GFP and its derivatives as labels for fluorescence microscopy (particularly confocal laser scanning fluorescence microscopy). Green Fluorescent Protein (GFP) has existed for more than 160 million years in one species of jellyfish, *Aequorea victoria*. In *Aequorea victoria*, a protein called aequorin releases blue light upon binding with calcium. The GFP then absorbs this blue light, giving off the green light. We can see when proteins are made and where they can go using GFP. The importance of GFP was recognised in 2008 when the Nobel Committee awarded Osamu Shimomura, Marty Chalfie, and

Roger Tsien the Chemistry Nobel Prize "for the discovery and development of the green fluorescent protein, GFP".

- In 1993–1996, super-resolution microscopy (SRM) was first demonstrated by Stefan Hell, who pioneered a new optical microscope technology that allowed capturing images with a higher resolution than was previously thought possible. This results in a wide array of high-resolution optical methodologies, collectively termed super-resolution microscopy. In recent years, computational and technological advances have afforded more excellent adaptations of established SRM techniques and new types of SRM development. Super-resolution imaging techniques rely on near-field (photon-tunnelling microscopy and those that utilise the Pendry Super lens and near-field or the far-field. SRM includes optical microscopic techniques that use advanced methods (physical and computation manipulations) to reduce the size of the PSF (point-spread function) and thereby resolve objects smaller than the diffraction limit of light. There has been an explosion of techniques capable of performing this function in the past 20 years. Additionally, combining SRM techniques with computer algorithms and faster processing speeds has further enhanced the narrowing of the PSF and the processing of large data sets commonly associated with SRM.
- In 2010, a research team led by Hong Zhou, UCLA professor of microbiology, immunology, and molecular genetics, reported in the journal "Cell" that they were able to see an individual atom of a virus and image the structure at such a high resolution at University College Los Angeles (UCLA). This was done using cryogenic electron microscopy (Cryo-EM), an electron microscopy technique applied on samples cooled to cryogenic temperatures and embedded in amorphous ice (Electron microscopy techniques will be discussed in Chap. 2). This was the first published instance of imaging biological complexes at such a resolution. They were able to image the viruses at 3.3 angstroms. An angstrom is the most minor recognised division of a chemical element (refer to

Table 1.1 from this chapter), around the distance between the two hydrogen atoms in a water molecule.

- In 2014, Two Americans and a German, Eric Betzig, Stefan Hell, and William Moerner, respectively, all working separately, jointly received the 2014 Nobel Prize in Chemistry for the development of super-resolved fluorescence microscopy, which allows microscopes to now "see" matter, smaller than 0.2 μm on optical microscopy. Optical microscopy had previously been held back by a presumed limitation that it would never obtain a better resolution than half the wavelength of light. However, the research projects completed by Stefan Hell, William Moerner, and Eric Betzig were able to bring optical microscopy into the nanodimension. Since their inventions, nanoscopy has allowed researchers to visualise the pathways inside living cells. They have seen how synapses move between nerve cells in the brain, track the proteins involved in neurodegenerative diseases, and follow proteins in fertilised eggs as they divide into embryos.
- In 2017, researchers from University College Los Angeles (UCLA) and Uppsala University further developed a small 3D-printed microscope called a mobile microscope, which fits a mobile phone and assists in the DNA sequencing of tumour cells and tissue samples. This mobile phone microscope records multi-mode images of the process sample into an algorithm, automatically examining the images to read the sequenced DNA bases. The microscope, which can capture multi-colour fluorescence and bright-field images, identifies the presence of targeted DNA sequences while examining the samples for tumours, bacteria, viruses, or fungal cells.
- In 2019, At the University of British Columbia, researchers were able to develop a laser microscope, infrared laser beams for microscopic melanomas with the potential to diagnose diseases like skin cancer and perform precision surgery without a single cut to the skin. This device specialises in a multi-photon excitation microscope using an ultrafast infrared laser beam that can image living tissue up to 1 mm in depth. The

microscope can treat the disease by intensifying the laser's heat instead of scanning living tissue. This device treats any structure in the body reached by light, including nerves or blood vessels in the skin, eyes, or brain.

- In 2020, engineers at the Massachusetts Institute of Technology developed dark-field quantum dots, which are small, mirrored chips that help produce dark-field microscopy images without needing dedicated expensive components. Slightly larger than a postage stamp and as thin as a credit card, the chip emits a hollow cone of light that can be used to generate detailed dark-field images of bacteria, as well as algae and similarly translucent tiny objects. Dark-field microscopy involves placing brightly lit samples against a black background to gain a clearer view of a sample with better contrast. The research team tested the chip using seawater and nonpathogenic strains of *E. coli* and placed each sample on the chip on a standard bright-field microscope platform. They produced clear, detailed images of individual microorganisms and bacterial cells, which were close to invisible under bright-field illumination.

Much change has happened in the field of microscopy since the late sixteenth century when father-son scientist duo Hans and Zacharias Janssen invented the world's first compound microscope. The technology of the microscope has developed from a simple light microscope to an electron microscope where the magnification has attained the level of a single atom. The significant properties and its use of light and electron microscope shall be discussed in the next chapter.

Key Learnings

- TEM produces a magnified image of a specimen. The 50–200-kV TEM electrons cannot penetrate biological specimens much thicker than 150 nm. This is where the ultrathin sections of a biological specimen need to be prepared to avail the penetration of an electron beam and achieve a magnified image of the specimen.

- Ultrathin sections are essential for ultrastructural studies of the biological samples and some material samples of polymers, membranes, etc., at nanometre-scale resolution.
- In 1986, Max Knoll and Ernst Ruska were awarded the Nobel Prize in Physics for developing the transmission electron microscopy technique.
- In ultramicrotomy, a biological specimen can be cut into ultra-thin sections on a nanometre scale, which can later be viewed to observe the fine internal structure of the specimen at the cellular level in TEM.
- To see the unseen world beyond the capacity of the human unaided eye led to the discovery of magnifying lenses, which were later associated with the development of optical lenses, which is the foundation of the compound microscope (Gr. Micros = small; skopein = to see) and various other microscopes in today's world.
- The first functional compound microscope was invented in 1590 by Francis Janssen and Zacharias Janssen, two Dutch spectacle-makers and a father-and-son team.
- The technology of the microscope has gradually developed from a simple light microscope to an electron microscope where the magnification has attained the level of a single atom.

Reference

Avers CJ (1978) Basic cell biology. Van Nostrand Reinhold Inc, New York

2 Light and Electron Microscope

This chapter explains light and electron microscopes, and under these two headings, their different components, functions, limitations, and technical approaches are also described. The standard protocol of biological sample preparation method for light and electron microscopy is also included to provide a general idea of specimen block preparation, which will facilitate the further processing of microtomy sectioning for light microscopy and UT sectioning with ultramicrotome for TEM study.

Learning Objectives from This Chapter

- Magnification and resolution of microscope
- Condenser lens and objective lens
- Light microscopy, its different types and standard protocol of biological sample preparation
- Electron microscopy, its different types and standard protocol of biological sample preparation
- Preparation and use of primary and secondary fixatives, buffers, embedding medium, and positive and negative stains

Both light and electron microscopes enhance our sense of sight and allow us to view things that are too small or invisible to view with the naked eye. Microscopes do it by magnifying (making

A. Haloi, *Essentials of Ultramicrotomy*,
https://doi.org/10.1007/978-981-97-4791-7_2

things appear bigger) and resolving (increasing the ability to distinguish between two points) the object.

The ability of a microscope to produce an image of an object at a larger scale than its actual size is magnification. This is done using a magnifying lens between the observer and the object, enabling a virtual image to form. The magnifying power is the extent to which the object is being viewed or appears enlarged, related to the geometry of the optical system. The magnifying power of a lens can be calculated by dividing the least distance of clear vision of an object by the distance from the lens to the plane at which the incoming light is focused, called the lens' focal length.

$$\text{Magnifying power} = \text{Lens with the least distance of precise vision} / \text{Focal length}$$

For example, a lens with a least distance of clear vision of 50 cm and a focal length of 5 cm (2 inches) will have a magnifying power of about 10×.

On the other hand, the shortest distance between two points of a specimen that can still be distinguished as separate entities is called resolution. The unaided human eye can distinguish two points 0.2 mm apart with sufficient light. This distance is called the resolution or resolving power of the human eye. If the points are less than 0.2 mm apart, it will be seen as only one point. Thus, the resolving power of the human eye is limited to 0.2 mm only. Resolution is related to the wavelength of light used to image a specimen; light of shorter wavelengths can resolve greater detail than longer wavelengths. Three mathematical concepts must be considered when dealing with resolution: Abbe's diffraction limit, Airy discs, and the Rayleigh criterion. Each of these is covered below in the chronological order.

Abbe's equation: The resolution of a microscope can be limited by some factors, like the use of imperfect lenses or a wrong alignment. However, there is a principal limit to the resolution of any optical system (i.e., eye or microscope), called diffraction limit. The limit of resolution (l) of any optical instrument is given approximately by Abbe's equation:

$$\text{Resolution}(l) = \text{Wavelength}(\lambda) / \text{Numerical aperture}(n \sin \alpha)$$

where λ is the wavelength (colour) of the illumination or radiation used to form the image, *n* is the refractive index (a function of density) of the material (i.e., mostly air or water) between the specimen and the first Lense (or objective Lense), and sin α is a sign of the semi-angle of the aperture of the first Lense as viewed from the specimen. The quantity "*n* sin α" is often called the numerical aperture (NA). Abbe's equation made it clear that high resolution in a microscope can achieved by adjusting the wavelength of the illuminating radiation, the refractive index, and the numerical aperture.

Airy disks: When light passes through an aperture of any size, diffraction occurs, as every lens has a finite aperture. The resultant is a bright region in the centre, with a series of concentric rings of decreasing intensity around it, known as Airy disks. Airy disk is also known as Airy disc and Airy pattern. The airy disk can also be defined as the best-focused spot of light that a perfect lens can make with a circular aperture and is limited by light's diffraction. The circular aperture creates a diffraction pattern of concentric rings that grow dimmer as we move away from the centre. An airy disk consists of a bright central spot surrounded by rings. The intensity of the first bright ring is 1.75% of that at the peak. The radius of the first dark ring of the Airy disk is

$$[1] R_0 = 0.61\lambda \, N \sin\alpha,$$

where λ is the wavelength, *N* is the refractive index of the immersion medium, and α is the angle subtended at the edge of the objective aperture. The radius depends on NA; for good resolution, it should be as large as possible. Thus, high-magnification lenses often use an immersion fluid, usually oil, with a refractive index of 1.518 (ISO 8036/1).

The diameter of this pattern is related to the wavelength (λ) of the illuminating light and the size of the circular aperture, which is essential since the Airy disk is the smallest point to which a beam of light can be focused. As focused Airy patterns from different object details approach one another, they overlap. When the overlapping patterns create enough constructive interference to reduce contrast, they become indistinguishable.

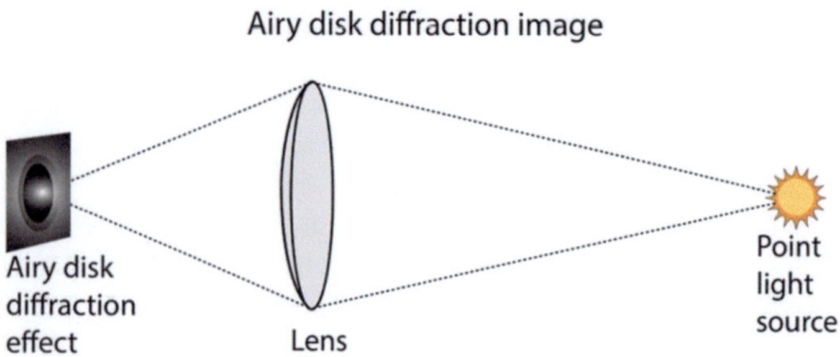

Fig. 2.1 Image of Airy disk diffraction

Because of the airy disks, point sources close to one another can overlap and produce a blurred image, as shown in the image (Fig 2.1).

Rayleigh criterion: As already discussed in resolution, it enables the separation of two closed objects or their images, which may look like one and make it hard for eyes to distinguish them, and the resolving power is the capacity of optical equipment to generate completely separate images of two extremely close objects. The inverse of the distance or angular separation between two objects that may be just about to resolve when viewed through a microscope is known as the Rayleigh Resolution Criterion. English physicist John William Strutt, 3rd Baron Rayleigh (1842–1919) investigated the Rayleigh criterion and is named after him (Fig. 2.2).

So, the Rayleigh criterion is defined as the minimum resolvable separation between the points, which is the radius of the Airy disk. The Rayleigh Criterion can be mathematically described using angular separation and single-slit diffraction. Angular separation can be calculated using the equation:

$$\theta = s / d,$$

where θ = angular separation, s = distance between the two sources, d = distance between the sources and the observer.

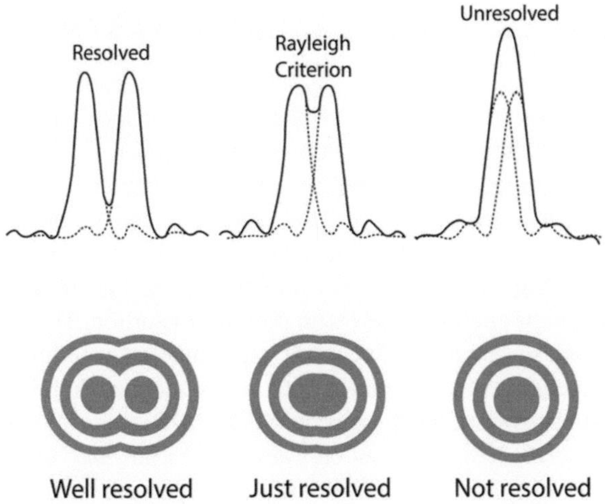

Fig. 2.2 Rayleigh criterion for resolution of objects

Angular separation, θ, is equal to the separation, s, of two sources divided by the distance, d, between the sources and the observer.

Another important fact is that the quality of imaging in microscopes is influenced by two essential components, i.e., the condenser lens and the objective lens. These two lenses work together to create clear and detailed images of specimens in microscopy. A microscope's magnification and resolution capacity depend on these two lenses.

The condenser lens, positioned below the stage of a light microscope (LM) and above the specimen in the case of an electron microscope, is responsible for focusing and directing light onto the specimen. The primary function of a condenser lens is to collect radiation (lights or electrons) from the illumination source and concentrate or condense it to a cone of light or electrons. Then, this focused beam passes through the specimen, illuminating it evenly to provide uniform and bright illumination and optimise image contrast. This makes the condenser lens crucial in achieving clear and detailed microscopic images. Also, the condenser lens assists in controlling the numerical aperture (NA) and the depth of field, enhancing the overall resolution and clarity of the observed specimen.

The objective lens, positioned above the specimen in the case of a light microscope and below the specimen in the case of TEM, captures the light that passes through the specimen and converts it into a magnified image. Objective lenses are obtainable in various magnification powers to facilitate scientists and researchers in observing the specimens at different levels of detail. The optical design of the objective lens is complex to correct aberrations and provide high-quality images with minimal distortion. The level of detail of a specimen is influenced by the numerical aperture of the objective lens, whose function is to determine the resolution and resolving power of the microscope (Fig. 2.3).

The critical differences between condenser and objectives lenses are,

The position of the condenser lens is below the specimen in the case of a light microscope and above the specimen in the case of an electron microscope. This lens collects and focuses the radiation onto the specimen in both cases. In contrast, the objective lens is positioned above the specimen in the case of a light microscope and below the specimen in the case of TEM, capturing the radiation that has passed through the specimen to form the magnified image of the specimen.

The condenser lens improves the image quality by improving image contrast and clarity, given uniform and bright illumination. On the other hand, the objective lens is responsible for magnify-

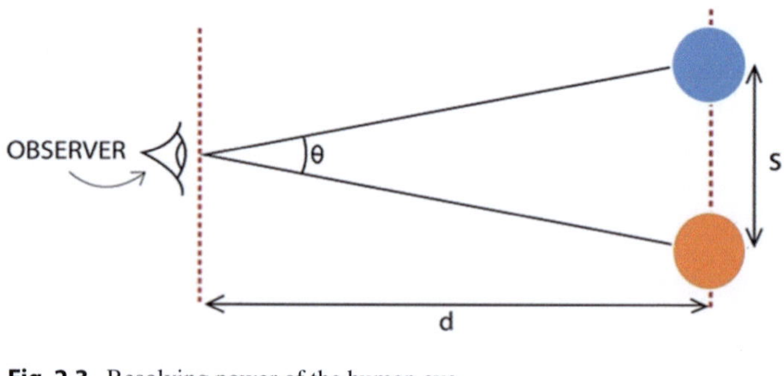

Fig. 2.3 Resolving power of the human eye

ing the specimen image and determining the level of detail observed.

A condenser lens does not contribute to the magnification of the specimen; on the other hand, an objective lens determines the level of magnification applied to the specimen, permitting detailed observation at different scales.

Understanding the different roles of the condenser and objective lenses, the resolution and quality of the image largely depend on the combined contributions of the condenser and objective lenses, making their synchronisation critical for obtaining accurate and detailed results in microscopy.

Light Microscopy

The LM is an instrument used to make small structures and samples visible by providing a magnified image of a specimen far beyond the limits of what we can see with our naked eye. The LM is used for visualising the fine detail of an object by creating a magnified image through a series of glass lenses, which first focus a beam of light onto or through an object, and convex objective lenses to enlarge the image formed. In most LMs, the image is viewed directly through binocular eyepieces that act as a secondary lens in the form of a magnifying glass to observe the magnified image. LMs are simple and compound. Simple microscopes comprise a single lens and are utilised at a fundamental level, where rigorous investigation is not required. It has a short focal length and a double convex lens. A simple microscope creates a virtual sample representation and produces a big image of a small item. The condenser lens is absent in a simple microscope, and its total magnification is limited to the magnification of the single lens used. Simple microscopes look at microscopic creatures like fungi, algae, and biological samples. Other than that, it is used to examine engravings, stamps, and other archaeological material in depth, in soil science and nutrition research and to research ancient texts and manuscripts. Also, it is used by watchmakers to examine the magnified image of small elements in the watch; Jewellers use it to examine and inspect jewellery; palmists and astrologers use it to examine an enlarged image of our fingerprints; Dermatologist use it to look for and diagnose skin allergies and disorders. A compound microscope uses numerous lenses to magnify tiny objects to the human naked eye and also serves as a basic microscope. A compound microscope has a flat mirror surface on one end. In contrast, at the other end, it has a concave mirror surface and is used for professional applications that require extensive investigation. They are commonly used in pathology labs to identify viruses and bacteria, to examine plant cells and the bacteria that live on them, and to determine the presence or absence of minerals and metals in blood samples and other biological materials. They are also used in a forensics lab to investigate crimes and conduct academic experiments in schools and colleges.

The LM uses visible light and a system of lenses to magnify images of small samples or specimens. The LM is often called a "light optical microscope" because it has two types of lenses to magnify an object, i.e., an ocular lens or eyepiece, which is close to the eye and an objective lens, which is close to the object. In this microscope category, a beam of light is directed through a thin object, and a combination of ocular and objective lenses generates the image of the object, which our eyes can view through an eyepiece. The image formed is realistic because it is a multicolour light. The LM can magnify the image of a specimen up to 40–2000 times. The recent development of LMs is known as super-resolution light microscopes (SRM), and it magnifies living biological cells up to 20,000 times or more.

Visible light has a wave-like nature with a 400–800-nm wavelength. The resolution of the LM is limited primarily by the wave nature of light, which means resolution is essentially linked to the numerical aperture (NA) of a microscope's optical components (objective lens, as well as the wavelength of light used). NA is a key factor of the objective lens (resolving power, focal depth, and brightness) for the performance of the microscope (Fig. 2.4). NA is also termed an object-side Aperture, a value defined initially by Abbe for microscope objectives and condensers. NA holds the position of important consideration and also a number without units, which is related to the refractive index (n) of a medium

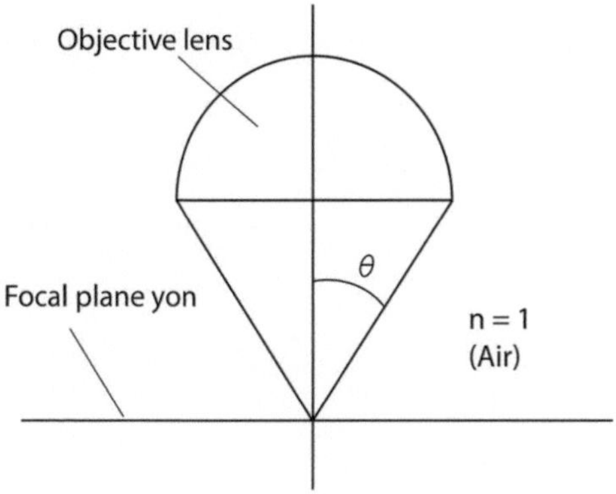

Fig. 2.4 Simple depiction of NA of an imaging system

through which light passes as well as the angular aperture (α) of a given objective when trying to distinguish detail in a specimen viewed under a microscope. To obtain a good resolution, the NA, or the measure of the solid angle covered by an objective, is defined as:

$$\text{Numerical Aperture}(\text{NA}) = n \times \sin(\mu) \text{or} \, n \times \sin(\alpha)$$

In the above equation, the variable μ or α stands for one-half angular aperture. In some instances, θ is also used for the same explanation, where n represents the refractive index (refractive index is the measure of bending of a light ray when passing from one medium to another. It is also defined as the ratio of the velocity of a light ray in an empty space to the velocity of light in a substance, $n = c/v$, where n = refractive index, c = speed of light, and v = phase velocity of light) of the immersion medium used between the objective front lens and the cover slip protecting the specimen ($n = 1$ for air; $n = 1.51$ for oil or glass) (Figs. 2.5 and 2.6).

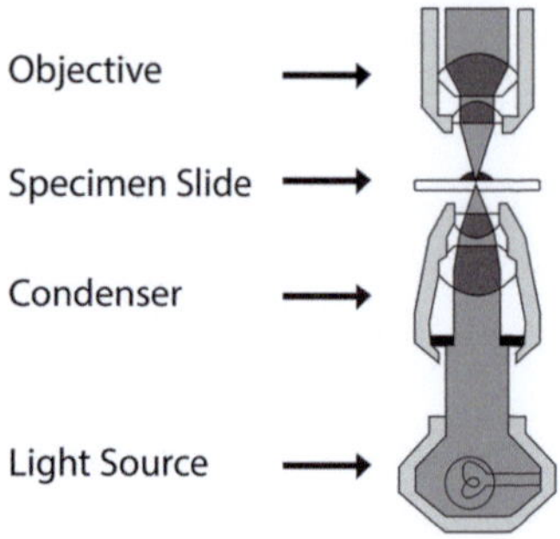

Fig. 2.5 Working principle of light microscope

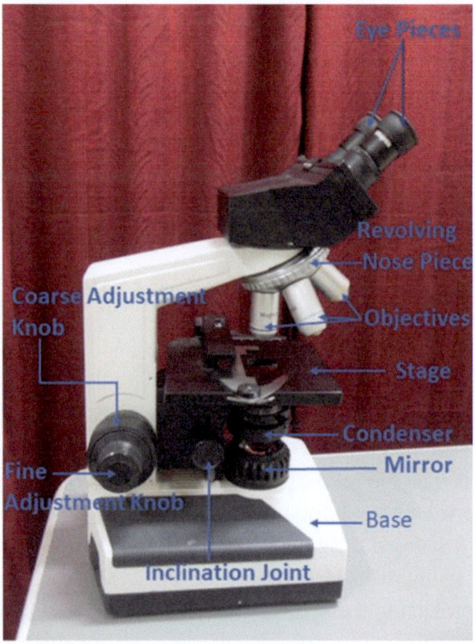

Fig. 2.6 The main components of a light microscope: (1) Eye Pieces, (2) Revolving Nose Piece, (3) Objectives, (4) Stage, (5) Condenser, (6) Mirror, (7) Fine and Coarse Adjustment Knob, (8) Inclination Joint, (9) Base

Different Types of Light Microscopes

To make full use of the available resolving power of the compound light microscopes, special techniques have been designed to improve contrast. Certain improved types of light microscopes are of the following types:

Darkfield Microscope

A Darkfield microscope is also called a dark-ground microscope and ultramicroscope. The darkfield microscope's principal work is to increase the contrast of the specimen viewed and only typically works with crowded or thick objects. It is used to study unstained biological sections or, if covered with silver or gold particles. A few common specimens for darkfield are cell cultures, microbes, suspensions of bacteria, plankton, foods, fibres, crystals, colloids, arthropods, auto radiographs or tissues labelled with metal particles and to examine pathogenic bacteria and in live blood analysis. It is a technique where the contrast in the specimen is increased, which is not imaged well under normal illumination conditions by oblique illumination since one sees only those light rays scattered from objects; the images appear bright on a black background (Fig. 2.7). In this technique, light passes through the specimen from oblique angles in the form of a hollow cone that is focused on the specimen, but only light that is

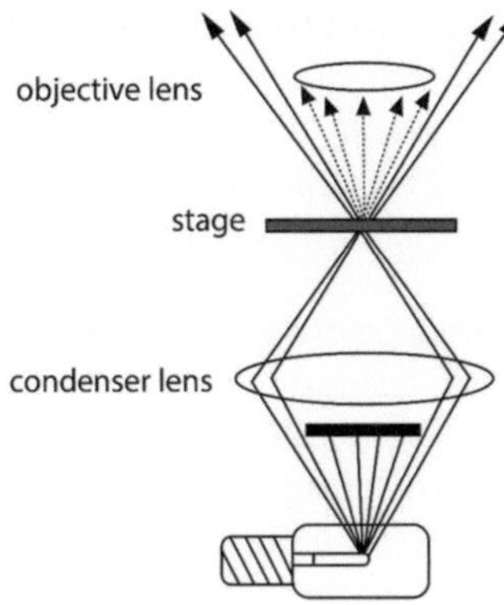

Fig. 2.7 Principle of dark field microscopy

reflected, refracted, or diffracted enters the objective to form a bright image of the specimen superimposed on a dark background after an opaque stop in the condenser has blocked the direct light. The specimens are generally highly refractile and must be spaced apart. This dark background provides a high degree of contrast, and the specimen appears bright against a black background and can produce striking images. Coloured filters or Rheinberg filters can also be used with this technique, as colours do not interfere with the actual information of the specimen. The darkfield microscopy method is used in light and electron microscopy fields, excluding the unscattered beam from the image. In the case of an optical microscope, a darkfield condenser lens should be used, which will direct the cone of light away from the objective lens. For a better result, the NA must be less than 1.0 of the objective lens and immersion oil must be used.

Brightfield microscopy cannot provide good results in some specific specimens as specimens are unstained or living, and their components have a refractive index close to their surrounding medium and are difficult to see under brightfield microscopes as the contrast is low. Some modifications are made to overcome these problems: dark field microscopy, which would prevent direct light from entering the objective lens, only light reflected or diffracted by the specimen is allowed to enter the objective lens. In darkfield microscopy, a dark field condenser that contains an opaque disc instead of the standard condenser is used. This disc blocks the light from entering the lens directly, and only the light reflected off from the specimen enters the objective lens. As there is no background light, the specimen appears light against a black background. Darkfield microscopy can be used to view blood cells, bacteria, algae, yeasts, shrimp, other invertebrates, diamonds, and other precious stones.

Phase Contrast Microscope

Phase contrast microscopy is a type of LM and was first described in 1934 by Dutch physicist Frits Zernike, where mainly the contrast of images of transparent and colourless specimens is enhanced. In LMs, different cell parts have different densities, so the refractive indices and the regions where the refractive index

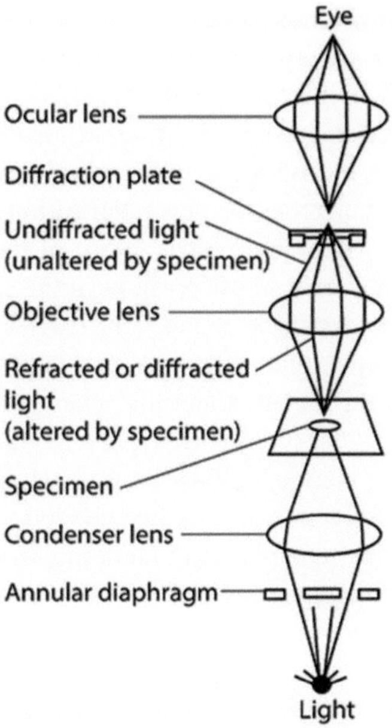

Fig. 2.8 Principle of phase contrast microscopy

changes tend to bend the light rays. These bent rays form destructive interference patterns and yield sharp contrasts between the specimen and the background. A phase contrast microscope is widely used to observe unstained and living cells, which enables the visualisation of cells and cell components that would be difficult to see using an ordinary LM (Fig. 2.8).

The significant advantages of phase contrast microscopy are that living cells can be examined naturally without being killed, fixed, and stained. So, the dynamics of ongoing biological processes can be observed and recorded in high contrast with sharp clarity of minute specimen detail.

Polarised Light Microscope

In this microscope category, polarised light is used to view specimens. In a polarised microscope, the specimen is placed between two close polarised and visible birefringent portions (*Birefringence* can be defined as the optical characters of a material having a refractive index that relies on the polarisation and propagation direction of light, Stein and Norris 1956) of the sample act like polarising film and hence, these portions of the sample are seen as bright objects on dark background. *Polarised light* is a contrast-enhancing technique that improves the quality of the image obtained with birefringent materials when compared to other techniques such as darkfield and brightfield illumination, differential interference contrast, phase contrast, Hoffman modulation contrast, and fluorescence. In this microscope, light vibrates in a specific plane rather than vibrating randomly in all directions like ordinary light. When light passes through certain types of materials, it can become polarised. This microscope uses this light property to reveal a specimen's internal structure and properties. It consists of a microscope body with an objective lens and eyepiece, a light source, and two polarising filters called polarisers. The polarisers are placed at the light source and the eyepiece, and they are oriented so that the planes of polarisation are perpendicular to each other. When a specimen is placed between the polarisers, it absorbs or reflects the polarised light differently depending on its properties, allowing the internal structure and properties of the specimen to be observed (Fig. 2.9).

This type of LM is useful mainly to observe minerals, crystals, and other transparent or semi-transparent materials as well as to analyse the structure and properties of these materials and highly ordered objects such as crystals or bundles of parallel filaments (i.e., microtubules or mitotic spindles).

Sample Preparation for Light Microscopy

Many sample preparation techniques have been developed in light microscopy depending upon the types of different specimens and

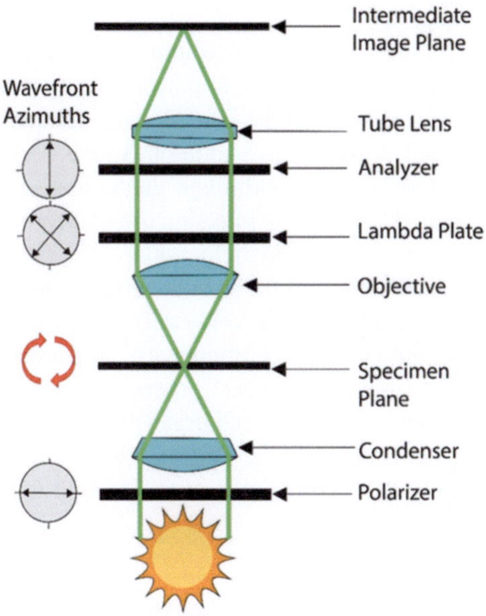

Fig. 2.9 Principle of polarised light microscopy

their wide range of applications. Some specimens are mounted on a microscope slide in a suitable liquid and then covered with a glass coverslip to examine under an LM. Visualising the living cells with light microscopy is only sometimes a good model, and so, for a better result, some additional steps are required. Protocols for sample preparation vary depending on the type of specimen, such as size and hardness, as well as expected post-processing, which includes planned staining techniques and the microscope used. Most sample preparation methods for light microscopy involve killing the cells or specimens. Light microscopy is typically used to study the cells and tissues in histology. The general sample preparation technique of studying histology for light microscopy is highlighted below.

Fixation

There are two types of specimen fixation techniques to preserve the cell organelles of the specimen in nearly live conditions, i.e., immersion and perfusion. In the immersion technique, the specimen is immersed in the fixative solution. In the perfusion technique, the fixative has to pass through the animal's bloodstream before removing the organs by injecting the fixative solution into the animal. This technique may help reduce artefacts and false or inaccurate representations of the specimen that result from chemical treatment or handling of the cells or tissues. Fixation kills the cells, stabilises their chemical components, and hardens the specimen in anticipation of further processing and sectioning. Chemical fixatives preserve the cell organelles by preventing degradation and help maintain the cell structure and sub-cellular components. The most common fixative for light microscopy is 10% neutral buffered formalin.

Dehydration

Most of the tissues are dehydrated by removing water from the specimen before the tissue is infiltrated with embedding medium or resin as water does not mix with embedding media, and this process is commonly carried out by immersing the specimen into a series of dehydrating agent in an increasing concentration till pure dehydrating agent is reached. Alcohol or acetone is miscible with water in all proportions. So, the dehydration process is carried out by treating the tissue with the dehydrating agent, commonly acetone or alcohol, continuously increasing concentrations of the dehydrating agent and decreasing water concentrations.

Embedding

As the specimen is tiny in size, the tissue needs support to be held by something solid, and this supporting medium is the embedding medium. Embedding medium will facilitate sectioning by holding the tissue rigidly in position while sections are cut. Through the Edging process, the specimens are enclosed in a mass of embedding medium using a mould. Various embedding mediums include paraffin wax, celloidin, synthetic resins, gelatine, carbowax, methacrylate, epoxy resin (Araldite), agar embedding, celloidin media, and gelatin. The choice of embedding media depends upon the tissue types, i.e., hard tissue like bone or soft tissue like liver biopsy, the type of microtome for sectioning the specimen and the type of microscope used for viewing the sample. Paraffin wax-embedded histological slides are the most popular for routine light microscopy, and the sections produced with this are called "paraffin sections". These sections are usually prepared with a "rotary" microtome. Different types of moulds are used to embed the sample depending upon the need, e.g., Leuckhard embedding moulds or L mould, paper blocks, plastic moulds, etc. The commonly used moulds are L moulds, made up of metal, easy to procure, reusable, and can be adjusted to make different sizes of blocks. One limb of the "L" is longer than the other. The two "Ls" are joined to form a side of the rectangular box that acts as a cast to make the mould. The sample is infiltrated by embedding media in a liquid state during the embedding process. After a period of cooling or polymerisation, the embedding media becomes solid without altering tissues' morphological or molecular features.

Sectioning

As the name suggests, the sectioning technique involves physically cutting the tissue to expose successively deeper parts of the specimen. This is done with a microtome, an apparatus for sectioning the blocks with an ultrasharp blade adjusted to micron-level precision. This process is done by mounting the paraffin or plastic block containing specimen on the arm of the microtome, which advances the block by small increments towards the metal or glass blade of the microtome. As successive sections are cut, they usually adhere to one another, forming a ribbon of thin sections. The sections are usually cut between 1 and 10 mm in thickness depending upon the type of sample and research question involved. As the sections are cut sufficiently, they are floated on a warm water bath to smooth any wrinkles. Sections are then picked onto a glass slide. The glass slides are heated in a warm oven or on a hot plate for about 15 min to help the section adhere to the slide. In some cases, this heating of slides may be avoided to preserve some characteristics such as antigenicity, and adhesive-coated slides may be substituted to pick up the sections in this case. Typical adhesives include starch, albumen, resins, and combinations. *Tissue Sectioning* is an art determined through experience with the type of tissue and the particular equipment. The selection of knife material, blade shape, cutting speed, knife angle, and other variables are firmly resolved with experience. Otherwise, tearing, ripping, venetian blinds, holes, folding, etc., may obtain.

Staining

The previous discussion establishes that light rays pass through the specimen to form an image in light microscopy. As the section of the specimen is transparent in its natural state, or in other words, most of the cell organelles and microorganisms lack colour and contrast, when the light rays pass through the section in a microscope, it does not produce an adequate image. This makes it difficult for microscopists to identify different cellular structures and their distinguishing characteristics. Stains help the specimen to enhance visualisation and give distinctive colour characteristics and contrast to different cellular components under a microscope. It is used to add contrast and label structures of interest. Common stains for light microscopy are eosin, methylene blue, toluidine blue, etc. Eosin stains the cytoplasm pink, while methylene blue stains DNA and RNA blue and toluidine blue stains different parts of the cell organelles with shades of blue. A series of stains or different stains can be used together to stain a specimen because each stain can colour differently to different cellular components. At the same time, a single stain is also used to stain the specimen in some cases. Numerous methods have been developed to assist the staining process.

Mounting of slide

After staining, the specimen containing the slide is ready for viewing in an LM. The slide is then positioned on the specimen stage of the microscope and examined through the ocular lens or with a camera. Magnification can be changed simply by rotating a turret with different objective lenses according to the requirements.

Limitations of the LM

Low Resolution: The resolution limitation in the LM is 200 nm as it uses normal white light, and the maximum magnification is 1500×. Any magnification beyond this is inadequate because of its relatively low resolution. So, an optical microscope cannot correctly view objects smaller than this. It means the optical microscope cannot be used to study smaller internal cellular structures like nuclei, ribosomes, endoplasmic reticulum, lysosomes, etc.

Electron Microscopy

Unlike the optical microscope, which uses visible light and a system of lenses to magnify images of small samples or specimens, the electron microscope (EM) uses a focused beam of electrons instead of photons and their wave-like characteristics to magnify an object's image. The advantage of an EM is that it can resolve features that are more than 1 million times smaller, which, in the case of LM, is between 5 and 2000 times. However, the recent development of LMs is known as "super-resolution" LMs and their magnification of living biological cells up to 20,000 times or more, as mentioned earlier in the light microscopy section of this chapter (Table 2.1).

As electrons cannot travel a noticeable distance in air at atmospheric pressure, a vacuum is required for the electron beam to pass through in an electron microscope. So, the column of the electron microscope is evacuated by pumps using airlocks. Thus, the specimen and the other necessary apparatus of the EM are introduced to a vacuum state. The electron microscope has variable-focus lenses, and the distance remains constant between the specimen and objective lens and the separation of the lenses. The magnification mainly depends on the value of the current (for magnetic lenses) through the intermediate and projector lens

Table 2.1 Comparison of light microscope (LM) to electron microscope (EM)

	LM	EM
Source of illumination	Light rays	Electrons
Wavelength	400–800 nm	2.5 pm (200 kV)
Medium	Air	Vacuum
Lenses	Glass	Electromagnetic
Image visibility	Direct	Fluorescent screen
Focusing/alignment	Mechanically	Electrically
Magnification	5–2000×	1,500,000 (200 kV)
Resolving power	200 nm	0.19 nm (200 kV)
Contrast	Absorption, reflection	Scattering, absorption

coils. The image can be focused by changing the current through the objective lens coil. The final image produced by the EM is invariably actual, which is visualised on a fluorescent screen and can be recorded on a photographic plate for study in a traditional EM and on a digital imaging system in the modern EM. The image in EMs results from the specimen's scattering of electrons by atoms. A heavy atom is more effective in scattering than one of low numbers, and the presence of heavy atoms in the specimen or UT sections of the specimen will increase the image contrast. Before viewing the specimen in an EM, the electron microscopist may incorporate or introduce more heavy atoms by staining the UT sections (which we will discuss later in Chap. 5). Modern instruments use electromagnetic lenses in place of electrostatic lenses in early microscopes.

All electron lenses show spherical aberration, distortion, coma, astigmatism, curvature of field, and chromatic aberration due to variations in the wavelengths within the electron beam, and these changes of electron velocity may be because of the variations in high-voltage supply to the electron gun or energy losses from collisions of electrons with atoms in the specimen. This difficulty may be resolved by carefully stabilising the high-voltage supply, and the high electron energies are commonly used for very thin specimens. The spherical aberration of the objective lens finally limits the microscope's resolving power. It is impossible to correct this aberration by adding a second lens of opposite characteristics, like in an optical microscope, because magnetic electron lenses are always convergent.

Astigmatism in an EM occurs when a magnetic lens is not perfectly round. The magnetisation of the metal in the lens changes every time switching on. The electron lens is mainly due to deviations from cylindrical symmetry in the radial components of the magnetic field of the lens. It is the result of an imperfect lens construction. It may usually be corrected using the stigma tors fitted to the objective lens.

The resolution is better if the wavelength is shorter, and the resolving power decreases as the wavelength increases. The resolution also improves with increases in the voltage because voltage reduces the wavelength.

In 1932, De Broglie defined the wavelength as:

$$\lambda = h / mv,$$

where h = Planck's constant, m = mass of the electron, v = accelerating voltage.

As the accelerating voltage (v) increases, the wavelength (λ) decreases; as a result, the resolving power also depends on various factors, such as the construction of electromagnetic lenses, etc.

Magnification is the ratio of image distance vs. object distance.

$$M = v / u,$$

where M = magnification, u = object distance, v = image distance.

Magnification can also be defined as the ratio of the resolving power of the eye to the resolving power (δ) of the microscope. The resolution cannot be less than half the wavelength used. Magnification has no meaning without the simultaneous increase in resolution.

$$M = \delta \text{ eye} / \delta \text{ microscope}$$

In 1926, Hans Busch, a stage director of opera who was associated with the Indiana University Opera Theatre for 32 years, invented the first electromagnetic lens, which was the principle behind the lens that invented the EM. In 1931, the first EM was developed by Ernst Ruska and Max Knoll, a physicist and an electrical engineer from the University of Berlin. This invention enabled us to see the world at an entirely different level. These two German scientists established a way to achieve a higher resolution than light. They found that accelerating electrons behave in a vacuum-like light to form an image of a specimen. They travel in straight lines and have about 100,000 times smaller wavelengths than light. Furthermore, it was found that electric and magnetic fields have the same effect on electrons as glass lenses and mirrors have on visible light. The discovery formed the basis of electron microscopy. This early model of Ruska and Knoll produced a magnification of 400 power and was the first device to examine the structure of samples with greater detail. This was the first type of microscope that had the potential to exceed the capabilities of the optical microscope. Afterwards, EMs were improved rapidly by many other scientists. In 1986, Ernst Ruska was awarded the Nobel Prize in Physics for the invention of the EM, in conjunction with Heinrich Rohrer and Gerd Binnig for the development of the scanning tunnelling microscope (STM). EM has become an indispensable tool and has significantly contributed to biological, biomedical, and material sciences. It is used for ultrastructural studies, diagnostic purposes, quantitative ultrastructural studies, nanoparticle characterisation, polymer structural features, diffraction patterns, etc. (Fig. 2.10).

In an EM, accelerated electrons carry a significant amount of kinetic energy, and this energy is dissipated as a variety of signals produced by electron sample interactions when the incident electrons are decelerated in the solid sample. In TEM, the electron beam passes through an UT section of the specimen (60–90 nm) or a film supporting the sample. The information about the sample is formed by the electrons that have passed through or around the sample, i.e., the transmitted electrons.

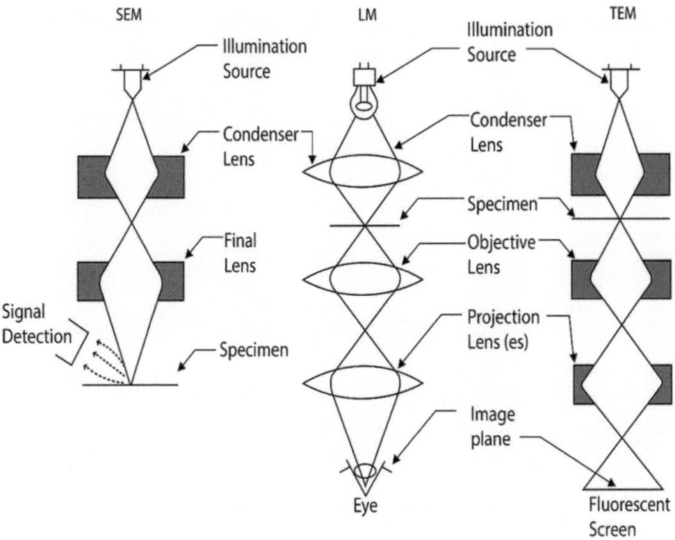

Fig. 2.10 Comparison of working principle among light microscope (LM), scanning electron microscope (SEM), and transmission electron microscope (TEM)

Different Types of Electron Microscopes

EMs have emerged as a powerful and essential tool for characterising various samples, including biological and material. EM has become a valuable tool for many applications because of its versatility and exceptionally high spatial resolution. The two main types of electron microscopes are the transmission electron microscope (TEM) and the scanning electron microscope (SEM) and another electron microscopy is a combination of TEM and SEM, called scanning transmission electron microscopy (STEM).

Although the TEM is the original and first EM, several other EMs are also developed for different purposes. The two major EMs are shortly summarised below:

Transmission Electron Microscope (TEM)

As the name suggests, a high-voltage electron beam transmits through the specimen to create a magnified image of the specimen, and the specimen is most often an ultrathin section of

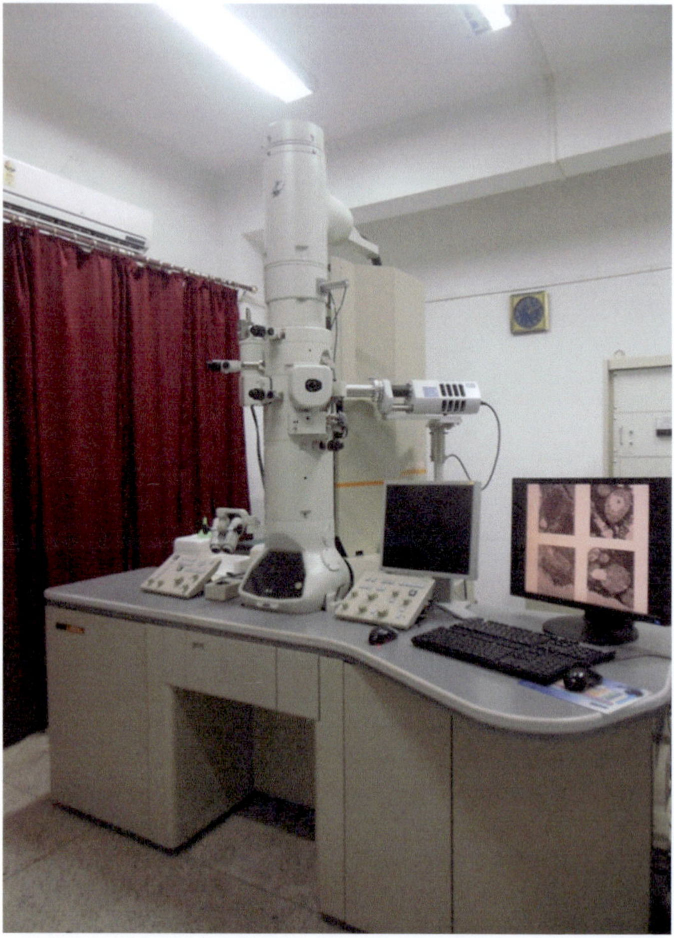

Fig. 2.11 Transmission electron microscope (TEM)

60–90 nm thick or a suspension on a grid. TEMs can magnify objects up to 2 million times (Fig. 2.11).

In TEM, an electron gun at the top emits electrons that travel through the microscope's vacuum tube. This gun is usually fitted with a tungsten filament cathode, producing the electron beam. An anode accelerates the electron beam, and electrostatic and electromagnetic lenses (glass lenses focusing the light in the case of LMs) focus the electrons into a very fine beam. This fine beam then passes through the specimen (UT section of the specimen), and the electrons either scatter or hit a fluorescent screen at the bottom of the microscope and produce an image of the microscopic structure of the specimen, which can be viewed through the objective lens of the microscope. An image of the specimen with its assorted parts shown in different shades according to density appears on the screen. This image can then be studied directly within the TEM or photographed. The spatial variation can be examined by projecting the image onto a fluorescent zinc sulphide-coated screen. The method to record the images of the specimen is to place a photographic film into the electron beam, which will record the image; a digital camera can also be used to display the image on a computer screen in real time. Traditionally, spherical aberration has limited the resolution of TEM. However, some developments have been made to overcome this issue to increase resolution with hardware correction of the spherical aberration. So, it is now possible to produce images with a resolution below 0.5 angstroms and more than 50 million times magnifications. The most significant limitation of the TEM is the requirement for fragile specimen samples, usually between 60 and 90 nm. This is a relatively lengthy and tedious procedure to achieve ultrathin sections of the sample (primarily biological), and there lies the need to chemically fix and dehydrate the specimen to be embedded in a polymer resin to make it possible to section and view the specimen details with a TEM. TEM can return an extraordinary variety of nanometre and atomic resolution information, in ideal cases revealing where all the atoms are, what kinds of atoms they are, and how they are bonded to each other. For this reason, TEM is regarded as an essential tool for nanoscience in biological and material fields. Max Knoll and Ernst Ruska demonstrated the first TEM in 1931, and 1986, Ruska was awarded the Nobel Prize in Physics for developing transmission electron microscopy.

Scanning Electron Microscope (SEM)

In the case of an SEM, electrons do not pass through the samples; instead, "Secondary Electrons" are reflected from the sample. The controlled movement of an electron beam is used to trace the object or specimen to create a magnified image of the original object on a monitor attached to SEM. The Electron Gun generates the electron beam and passes through the Condenser Lens, which condenses the beam onto the Objective Aperture of SEM and then further focuses the beam into a probe. To manipulate the electron beam, SEM uses scanning coils, which can move the beam precisely back and forth and create a magnetic field using fluctuating voltage. As the electron beam traces the topography and composition of the original specimen, it interacts with the surface of the specimen. It emits secondary electrons, which are responsible for image formation and other signals from the surface of the specimen. A secondary electron detector attracts those scattered electrons and, depending on the number of electrons that reach the detector, records different brightness levels on a SEM monitor. An additional sensor detects the electrons that reflect off the speci-

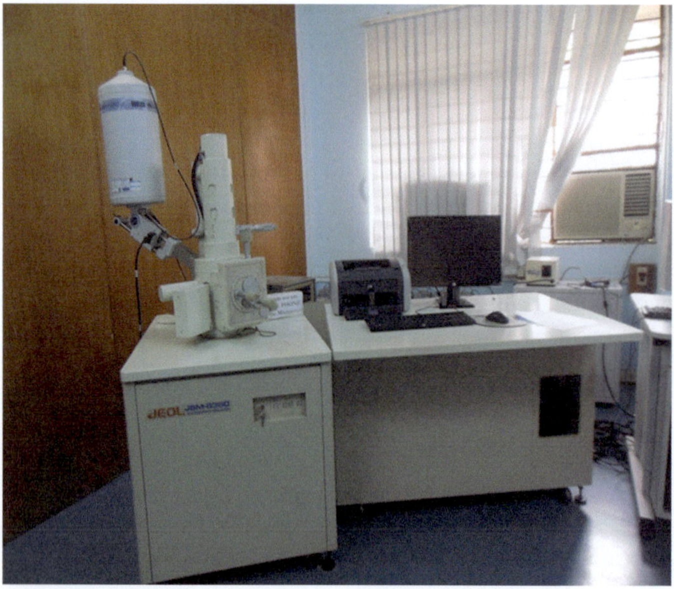

Fig. 2.12 Scanning electron microscope (SEM)

men's surface, called back-scattered electrons, and the electrons emitted from beneath the specimen's surface, called X-rays. An image of the original object in detail is scanned onto a monitor for viewing, as the name of the instrument "scanning" is. The thickness of the sample is immaterial in scanning electron microscopy (SEM) (Fig. 2.12). However, it is beneficial because it utilises surface processes, can create images of large samples up to several centimetres in size, and has a greater depth of field. As a result, the pictures from an SEM can be good representations of the natural shape of the specimen.

Sample Preparation for Electron Microscopy

Living biological samples cannot be directly viewed on EMs. So, the samples must undergo a complex preparation method, which involves many steps to help the specimen withstand the environment inside the microscope. It is a challenge for the scientist to preserve the specimen information to look as accurate as possible, like it would in living conditions, while still being able to withstand being visualised in the EM.

The reasons why living things cannot survive in an electron microscope are:

- The power of the electron beam that's directed at the specimen: The electron beam inside a transmission electron microscope (TEM) produces high energy, generally up to 150 °C, to pass through the ultrathin section of the specimen where the beam hits the sample, which is too high for living cells to survive. Although Scanning electron microscopes (SEMs) use a lower-energy electron beam than transmission electron microscopes (TEMs), they can still damage the specimen.
- The vacuum inside the microscope: Biological samples contain abundant levels of water, which evaporates immediately in a vacuum condition, and on the other side, electrons cannot travel without a vacuum. Instead, they would be deflected when hit by air particles. So, it would not be possible for the

emitted electrons to pass through the specimen to generate an image.

Sample preparation for electron microscopy before the viewing and analysis in EM is critical and directly impacts the quality of the characterisation of a sample. The basic philosophy behind specimen preparation is to manipulate the specimen as little as possible, ensuring that the sample is of appropriate size, stable in the vacuum, electrically conductive and possessing characteristics resembling the natural state. All through the sample preparation process, a sample must meet specific criteria, these are:

- Complete lack of water or other volatile components.
- Ability to remain unchanged under high vacuum conditions.
- Stability when exposed to electron beam damage (thermal and physical stress).
- Regions of both electron opacity and electron transparency (TEM).
- Appropriate size for the EM.

Natural topographies of most biological specimens will not be retained in a high vacuum. So, to ensure proper imaging and reliable analysis of physical specimens in an electron microscope, the specimens need to be fixed (stabilised), cleaned, and dried prior to analysis. As a result, the electron beam does not destroy them, and the vacuum does not affect them. The softer and more hydrated the sample is, the more care must be taken in its preparation. The specimen, when collected, should be stabilised immediately. Using gloves in every sample preparation step and loading into the microscope is mandatory. The steps of sample preparation for EM are as follows:

EM sample preparation depends mainly on the sample characteristics, and it may need some or all of the following steps,

Fixation

Fixation is the first and perhaps the most crucial step of biological specimen preparation in EM. In fixation, living tissue is chemically treated, which kills the living organism's tissue to stabilise. The fixation should be followed up immediately after dissecting the tissue. Once the tissue is removed from its natural environment, it starts to alter because the oxygen levels begin to drop as soon as it is removed from an organism. Hence, the look of mitochondria starts to modify, and the lipids form micelles. The purpose of fixation is to bring about the rapid cessation of biological activity of tissue and to preserve the structure of the cells, thus protecting them against alteration during the successive preparation steps like embedding and sectioning and preparing them for subsequent treatments such as staining and exposure to the electron beam.

Commonly used fixatives are Glutaraldehyde (C5H8O2), Paraformaldehyde (CH2O), Acrolein (C3H5O), Karnovsky's fixative (Glutaraldehyde+Paraformaldehyde), and Osmium tetroxide (OsO_4).

The most commonly used primary fixative is Karnovsky's fixative and 2–3% Glutaraldehyde made in 0.1M sodium phosphate buffer (pH 7.4). In most cases, primary fixation is done by immersion method, which is carried out by immersing the sample in the fixative immediately after dissection. Since the ability of penetration of the fixative is poor, the size of the tissue has to be very small, 1–2 mm^2 and is carried out at 4 °C. Timing for fixation of sample varies with different types of sample. For soft tissues, e.g., liver, kidney, spleen, ovaries, testis, etc., it is usually 2–4 h. Hard tissues like skin, scales, roots, stems, etc., are generally fixed for 12–24 h and cell suspensions such as blood, bacteria, and D.L. cells are generally set for 1–2 h.

Karnovsky's fixative has proved to be superior to any other aldehyde used alone and is the most commonly used primary fixative, followed by the secondary fixative, osmium tetroxide. Aldehydes fix protein and carbohydrates during primary fixation due to their cross-linking ability. The samples can be fixed by two processes, which are perfusion, where the fixative is introduced via the vascular system and immersion, where the specimen needs to be dissected to 1 mm cube size in at least one direction. One percent Osmium Tetroxide (OsO_4) solution is the most frequently used secondary fixative and is generally used after Karnovsky's fixation. OsO_4 reacts principally with lipids and also acts as an electron-dense stain. During secondary fixation with osmium tetroxide (OsO_4), it fixes the lipid portion of the tissue as the phospholipids forming membranes are preserved and are not extracted during dehydration. OsO_4 fixation forms a black insoluble precipitate on the membranes, creating contrast.

The affinity of different fixatives to react with various chemical constituents of the cell varies; hence, to obtain a complete picture, several fixation procedures are followed. Apart from aldehydes and osmium tetroxide, other compounds can partly fix and partly stain cellular constituents, providing additional contrast called tertiary fixation, including chromium salts, uranium salts, lead compounds, and phosphotangstic acid (PTA).

Preparation of Karnovsky's Fixative (Primary Fixative)

- Take 20gm of paraformaldehyde in 250 ml of 0.2M sodium cacodylate buffer or 0.1M phosphate buffer and heat to 600C.
- To dissolve the paraformaldehyde (drops of 1N Sodium hydroxide solution can be added in case of non-dissolving paraformaldehyde). Cool to room temperature.
- Add 20 ml of 25% Glutaraldehyde and 12.5gm of anhydrous calcium chloride.
- Add double distilled water to make the total volume to 500 ml.

Preparation of 3% Glutaraldehyde Fixative (Primary Fixative)

- Take 30 ml of 50% glutaraldehyde (or 60 ml of 25%) solution and mix the volume to 250 ml with double distilled water.
- Add 250 ml of 0.1M phosphate or 0.2M Cacodylate buffer.
- Add 12.5 mg of anhydrous calcium chloride (only when cacodylate buffer is used).
- Store at 40 °C.

Preparation of Osmium Tetroxide (Secondary Fixative)

OsO_4 is usually supplied in 1gm ampoules. A stock solution of 2%OsO_4 is prepared by dissolving 1 g in 50 ml of doubled distilled water. OsO_4 needs about 24 h to dissolve. It is stored at 4 °C in a tightly stopped brown-coloured bottle.

A part of 2%OsO_4 solution and an equal part of 0.2M sodium cacodylate buffer or 0.1M phosphate buffer is added to achieve

1% OsO_4 in 0.1M buffer solution to prepare the working solution, which is always prepared fresh before use.

Buffers

The fixing agents may have erratic effects on the permeability of cell membranes and, consequently, on cell organelles. If used alone, fixatives can change the molarity, molality, osmolarity, P^H, etc., of the tissue fluid; hence, exposing tissues directly to fixatives is not advisable. To overcome these problems, some buffers act as a fixative vehicle. The buffer serves as a solvent for the fixative constituent and also helps maintain a specified P^H, molarity, molality, osmolarity, etc., of the cell. The most commonly used buffers are phosphate buffer and cacodylate buffer.

Phosphate buffer (0.1M) is used with the fixative for the fixation of plant tissues as the best vehicle. It does not allow the extraction of cellular materials, especially amino acids and proteins. A phosphorate buffer is more efficient than a cacodylate buffer during the immersion fixative method in facilitating the penetration by the fixative.

Cacodylate buffer is the second most commonly used buffer for electron microscopy. It lacks the strongly reactive character of phosphate solutions. It thus can be used in cytochemical reaction mixtures and with media containing various ions without any chance of precipitation with other solution components.

Preparation of 0.1M Phosphate Buffer

- Sodium dihydrogen phosphate ($NaH_2PO_4 \cdot H_2O$)—27.6 g, sodium dihydrogen phosphate ($NaH_2PO_4 \cdot 2H_2O$)—31.21 g.
- Add double distilled water to make the volume to 1000 ml.
- Disodium hydrogen phosphate ($Na_2HPO_4 \cdot 7H_2O$)—53.65 g, or Disodium hydrogen phosphate ($Na_2HPO_4 \cdot 12H_2O$)—71.64 g.
- Add double distilled water to make the volume 1000 ml.

Add 36.0 ml of the second solution to 14.0 ml of the first solution and dilute to 100 ml with double distilled water. The solution thus prepared is 0.1M with a pH of 7.2.

Preparation of 0.2M Sodium Cacodylate Buffer

- Take 21.4 g of sodium cacodylate [$Na(CH_3)_2AsO_2 \cdot 3H_2O$] in 250 ml double distilled water.
- Add 3.45 ml of 1N HCl (0.83 ml of concentrated HCl in 10 ml of double distilled water).
- Add double distilled water to make up the volume to 500 ml.

Washing

Washing of the specimen is done every time after primary and secondary fixation. After primary fixation, specimens should be rinsed thoroughly with buffer (same as used for preparing fixative) to wash off the excess fixative for 15 min (three changes) at 4 °C (refer to "Standard protocol for specimen processing" specified later in this chapter). If traces of the primary fixative are left in the specimen, they may react with the secondary fixative (osmium tetroxide) and produce a precipitate. This is followed by Secondary fixation or post-fixation. After post-fixation, the specimen should be rinsed thoroughly with buffer (same as used for preparing fixative) to wash off the excess fixative.

Dehydration

Biological tissues contain about 75% water. It must be replaced with suitable organic solvent(s) before infiltration when water-immiscible resins are used for embedding. Complete water removal is unnecessary if the tissues are embedded in water-miscible resins. The water is removed by passing the tissues through a series of solutions of ascending concentration of the dehydrating agent. A fixed specimen is dehydrated by incubation in a series of ethanol or acetone solutions as dehydrating agents. Solvent concentration is gradually increased so water is removed gently without causing artefacts, mainly shrinkage. Acetone has an advantage over ethanol as it is more easily miscible with clearing agents, viz., xylene and toluene. Dry acetone is generally used, prepared by adding sufficient anhydrous copper sulphate to absolute acetone. The solution has to shake well, leave overnight to stand, and be filtered before use.

The above points are common for SEM and TEM sample preparation but differ from this step onwards. The following steps have to be followed for SEM sample preparation after dehydration.

Drying

After the specimen is dehydrated, it is dried using one of the various methods appropriate to the sample type. Drying the sample is essential because water vaporisation will obstruct the electron beam and reduce the clarity of the micrograph in a high vacuum environment of SEM. Usual drying methods for SEM include air drying, critical point drying (CPD) and hexamethyldisilazane (HMDS) or tetramethylsilane (TMS) treatment. Samples like egg shells, tree bark, bones, etc., can be air-dried. Other drying protocols are:

Critical Point Drying

Increasing the pressure and temperature of a wet sample makes it possible to dry the sample by achieving the "critical point" of that liquid at which the phase boundary between its liquid and gaseous states is eliminated. The gas is released slowly at this state to minimise the effect of surface tension forces on cell ultrastructures. The critical point for water is exceptionally high at 228.5 bar (3212 psi) and 374 °C, which could instantly destroy the samples, and the critical point for ethanol is 60.8 bar (882 psi) and 241 °C. Because of these properties of water and alcohol, they are unsuitable for the required point drying method.

Liquid carbon dioxide (CO2) with a critical point at 73.8 bar (1070 psi) and 31 °C is generally used for this method. Since CO_2 is not miscible with water, acetone is a transitional (intermediate) fluid. Samples are dried with liquid CO_2 at its critical point, i.e., 31.5 °C at 1100 psi, with an instrument called a critical point drier (CPD) in which the specimens and the dehydrating agent are introduced. The Liquid CO_2 gradually replaces the dehydrating agent. Once the exchange is complete, CO_2 is brought to a little above its critical point; the liquid CO_2 changes to vapour without a change of density and, therefore, without surface tension effects, which distort morphology and ultrastructure. The exchange and release of the CO_2 liquid and gas should be controlled to avoid turbulence within the CPD, which may be detrimental to the sample structures.

Hexamethyldisilazane (HMDS) or Tetramethylsilane (TMS) method (Dey et al. 1989): In this method, the specimens are immersed in TMS for 5–10 for two changes at 4 °C, and then they are brought to room temperature (25–26 °C) to dry. TMS, unlike acetone or alcohol, having a lower boiling point and vapour pressure, exits the sample with minimal damage or distortion to the sample.

Mounting of the Specimen

The dried specimens are then mounted onto aluminium or brass stubs using an adhesive paint or double-sided conductive tape. Sample orientation, sample stability, and ensuring good sample adherence on the SEM stub are some factors that need to be considered in the sample mounting process. A low-power stereomicroscope is recommended during the process to protect the sample area from a simple unintentional graze with a tool.

Coating

Biological samples, because of their non-conductivity, give rise to charging problems in the SEM as the bulk of primary electrons from the electron beam remain in the sample to form clouds of negative charges, and this built-up charge *interferes with the primary beam* to bring about image distortions, loss of contrast with very bright and dark areas, known as "charging effect". Therefore, its surface needs to be coated with a thin metal film so that the surface has conductivity. A *thin layer of metal* is sputtered on the sample, thereby *increasing its conductance* to enable the absorbed electrons to find their way to the ground. A sputter coater is used for this purpose. This technique is called *coating* and is achieved by using Ion sputters. In general, the principle used in the ion-sputter coater is that positive ions, produced by discharge in a low vacuum of about 10 Pa, sputter a target metal, and the specimen is coated with the sputtered target metal. Since the inside of this device is kept at a low vacuum, the particles of the sputtered target metal collide with many residual gas molecules; therefore, the metal particles are strongly scattered and attached to the specimen surface from every direction, forming a uniform film on the surface. The preferred metals for sputtering are gold, gold–palladium, platinum, and chromium. However, platinum and chromium are the choice metals for high-resolution imaging (more than 50,000×), as gold sputtering results in visible graininess on the sample's surface. Other materials used for sputtering are iridium and carbon (Bozzola and Russel 1999; Dykstra 1992; Hayat 1984).

The dry, mounted, and coated sample is now ready for examination under the SEM.

In the case of TEM sample preparation, the following steps are to be followed after dehydration:

Clearing

It is advantageous to clear off the dehydrating agent with a clearing agent to facilitate resin penetration before the infiltration step, even though acetone is readily miscible with resin. In their pure form, xylene, toluene, or epoxy propane are considered good clearing agents. Toluene is very effective, as it can remove any traces of water in case of incomplete dehydration.

Resin Infiltration

The infiltration step is done by gradually decreasing the concentration of the clearing agent and proportionately increasing the concentration of the embedding medium of liquid resin (typically epoxy resin) for ultrastructural studies. Very few tissues are rigid enough to be cut into thin sections without additional support. In the hard tissues, infiltration is carried out under vacuum as slow diffusion of the embedding medium is problematic. For plant tissues, vacuum treatment is routinely employed. This treatment also eliminates the air bubbles attached to the tissue. Infiltration is carried out with liquid resins, with which tissue embedding is carried out. The resins are of different kinds used to meet specific requirements. The most commonly used resins are Araldyte CY212 (high viscosity resin) and Spurr (low viscosity resin).

Embedding

Embedding is done in resin using a flat embedding mould or beam capsule. The flat embedding mould is used when a specific orientation is needed for a particular specimen. After filling the capsule or flat embedding mould with the resin, tissue is put in and ensures that it settles at the tip (bottom when placed vertically) and is filled with extra resin if required. Paper label written with pencil is also put along the side of the wall of the capsule or in flat embedding mould.

Three primary embedding mediums are used for embedding a tissue, i.e., paraffin wax, optimal cutting temperature (OCT), and resin. Each of these has its advantages and disadvantages, and the use of it depends on the research question and plan involved in it. Choosing which embedding medium to use depends upon which sectioning technique is engaged for the particular research. Some embedding mediums can be used for more than one purpose (Table 2.2).

The commonly used embedding medium for electron microscopy is Araldite CY212 and Spurr.

Preparation of Araldite CY212:

• Araldite CY212	7.5 ml
• Hardener DDSA (Dodecenyl Succinic Anhydride)	7.5 ml
• Accelerator DMP 30 (2,4,6Tridimethylamino methyl phenol)	300 μl
• Plasticiser (Dibutyl phthalate)	700 μl

Preparation of Spurr:

• Epoxy Resin ERL 4206	10.0 g
• Hardener NSA	26.0 g
• Reactive flexibility DER 736	6.0 g
• Accelerator DMAE	0.4 g

Add the above gradients and stir vigorously to mix them thoroughly. The complete medium must always be used in a fume cupboard as the volatile vapours may be harmful. Allow the air bubbles to settle down before use. For the air bubbles to settle, the mixture may be kept in an incubator in low heat and checked often

Table 2.2 Three primary embedding mediums and their uses

	Paraffin wax	OCT	Resin
Uses	Light microscopy	Light microscopy and Western blot	Electron microscopy
Fixation	Perfusion or immersion fixation in 4% paraformaldehyde	Snap freezing	Perfusion or immersion fixation in 4% paraformaldehyde or Karnovsky's fixative and post-fixation with osmium tetroxide (OsO_2)
Equipment used for sectioning	Microtome	Cryostat microtome	Ultramicrotome
Section thickness	4–5 μm	1–100 μm	<1 μm
Advantage	Well, the balance between good preservation and antigenicity.	Tissue fixation is not required, and tissue processing is minimal.	Because of ultrathin sectioning, the high morphological characteristics are preserved.
Disadvantage	Cutting of thin sections is difficult.	Sections are delicate	Ultrathin sectioning is a tedious and time-consuming process, and there are lots of toxic chemicals involved.

until the air bubbles are obliterated. Leaving the mixture in the incubator for a longer time and the high heat may cause polymerisation of the medium.

Polymerisation

Polymerisation is done by infiltrating resin into the tissues and further hardening it. Thus, the specimen is placed in a mould filled with liquid resin and cured into a hard block using heat or UV light that will take the form of a block embedded with tissue. The process should be carried out slowly as faster processing may cause blocks to polymerise non-homogenously and may lead to artefacts within the specimen. Another reason is that at 40–50 °C, most embedding media become less viscous, facilitating maximum penetration into the tissue. For this reason, the specimen is kept overnight at 40–50 °C and then increased to 60 °C so that the resin gets hardened with an incubator with reasonable temperature control for this purpose. After the block is prepared, a sample can be stored indefinitely.

Ultrathin (UT) Sections

After the preparation of blocks, UT sections are cut using fresh glass knives or a diamond knife with an ultramicrotome. UT sections show interference colours while floating on the liquid of the trough. This makes it possible to determine the thickness of the sections.

Grey—60 nm, optimal for high-resolution work.
Silver—60 to 90 nm, ideal for most purposes.
Gold—90 to 150 nm, useful for low magnification.

The UT sections floating on the water are stretched by exposing them to chloroform vapours. The sections are lifted from below on specially made grids. Grids made of copper are commonly used. Grids made of gold, platinum, nickel, and stainless steel are also available and are used for specific purposes.

The grids are available as meshed or holed; grids of 100–300 mesh size are most commonly used. One of the surfaces of the grid is shiny, and the other is matted. The sections are lifted using the matted surfaces of the grid, which allows the sections to be stuck firmly.

If grids of large mesh size or holed are used, a film coating is essential to render support to the sections. Film coating is done on the matted surface of the grid. 3% formvar (Polyvinyl formaldehyde) in ethylene dichloride is generally used for coating. The Coated grids help support particulate specimens like viruses and bacteria. The ultramicrotomy procedure will be discussed briefly in Chap. 4.

Staining of UT Sections

- Positive staining: A double staining method using uranyl acetate and lead citrate is routinely followed to obtain a good contrast. The prepared stains need to be filtered and centrifuged before every use. A small amount of *uranyl acetate* solution should be pipette out in a clean parafilm sheet. The grid containing ultrathin sections is immersed into the stain with the sample side up for 30–40 min at around 30 °C, and a wooden or cardboard cover has to be placed over it as the staining is effective when carried out in the dark. The excess stains are then washed off in running double distilled water and dried on a filter paper with the section side up again.

The uranyl acetate stained grid carrying the section was then placed onto a few ml of *Lead citrate* stain on a parafilm sheet for 5–10 min at around 30 °C. In this case, the grid has to float on the stain with the sample side down. Excess stain is washed off in running double distilled water and dried the grids on a filter paper, sample side up. Now, the grids can be stored in grid boxes or a well-covered petri dish. The sections are ready for observation in the TEM.

Preparation of Stains for Positive Staining

Uranyl Acetate

Prepare a saturated solution by adding an excess uranyl acetate (annular) to 10 ml of filtered 50% ethanol in a 15 ml centrifuge tube. Shake vigorously for 2 min and spin down hard to allow the excess of uranyl acetate to settle. The solution is ready for use. Stopper and store at 4 °C. This stain is always prepared fresh before use.

Lead Citrate

Add one-half of a pellet of sodium hydroxide to 12 ml of double distilled water in a 15 ml centrifuge tube. Shake well to dissolve. Add 50 mg of Lead citrate (analar), shake well for 2 min, and centrifuge. The solution is ready for use. Stopper and store at 4 °C. This is always prepared fresh before use.

- Negative staining: To examine fine surface details of small particulate specimens, "negative staining" has proved to be an important technique. The method consists of embedding suitably small particulate specimens in a layer of dense material so that the specimen is viewed as a light object against a dark background.

A small objective aperture should be used to view the negatively stained specimen under the TEM. One should operate the microscope at the lowest practical accelerating voltage if one wants maximum contrast from negatively stained material.

Preparation of Particulate—Specimens for Negative Staining

1.0 to 3.0% solutions of phosphotungstic acid (PTA), ammonium molybdate, or uranyl acetate are used as a negative stain and of these, PTA is the most commonly used stain.

Staining Procedure

- Prepare 2% PTA in double distilled water or 0.1 to 1.0% ammonium acetate.
- Adjust the pH from 6.8 to 7.4 with 1N potassium hydroxide (KOH). The resultant Potassium Phosphotungstate forms an electron-dense background when dried.
- Place a drop of the suspension of the particulate specimen (in higher concentration) on a coated grid and drain off the excess. Do not allow it to dry.
- Add a drop of PTA and allow to stand for 2 min. Drain off the excess.
- Dry and observe under the TEM.
- The unstained minute details will be seen as electron-lucent structures in an electron-dense background.

Standard Protocol for Specimen Processing for SEM

Standard protocols for biological samples (2–3 mm cube in size) are as follows,

- Primary fixation: 2.5–3% Glutaraldehyde or Karnovsky's Fixative.
- Washing: Washed in 0.1M sodium cacodylate buffer 3 changes of 15 min each at 4 °C.
- Post-fixation (Optional): 1% osmium tetroxide in 0.1M sodium cacodylate buffer at 4 °C.
- Washing: Washed in 0.1M sodium cacodylate buffer 3 changes of 15 min each at 4 °C.
- Dehydration: All steps to be carried at 4 °C.

30% Acetone 15 min × 2 changes
50% Acetone 15 min × 2 changes
70% Acetone 15 min × 2 changes
80% Acetone 15 min × 2 changes
90% Acetone 15 min × 2 changes
95% Acetone 15 min × 2 changes
100% Acetone (dry acetone) 15 min × 2 changes

- Drying:

Critical Point drying: Samples will be dried with liquid carbon dioxide at their critical point, i.e., 31.5 °C at 1100 p.s.i.

TMS method: The specimens are immersed in tetramethylsilane for 5–10 for two changes at 4 °C. Then, they are brought to room temperature (25–26 °C) to dry.

- Mounting: The specimen is mounted on brass or aluminium stubs.

- Sputter coating: Coating uses silver or gold; the layer should be about 35 nm thick.

Standard Protocol for Specimen Processing for TEM

Standard protocol for biological samples (1–2 mm cubed sized),

- Primary fixation: 2.5–3% Glutaraldehyde or Karnovsky's Fixative (Karnovsky 1965) for about 1–24 hour, depending upon the hardness of samples at 4 °C.
- Washing: Wash in 0.1M sodium cacodylate buffer—3 changes of 15 min each at 4 °C.
- Post-fixation: 1% osmium tetroxide in 0.1M sodium cacodylate buffer at 4 °C for about 1 h.
- Washing: Wash in 0.1M sodium cacodylate buffer 3 changes of 15 min each at 4 °C.
- Dehydration: All steps are to be carried out at 4 °C.

  30% Acetone 15 min × 2 changes
  50% Acetone 15 min × 2 changes
  70% Acetone 15 min × 2 changes
  80% Acetone 15 min × 2 changes
  90% Acetone 15 min × 2 changes
  95% Acetone 15 min × 2 changes
  100% Acetone (dry Acetone) 15 min × 2 changes
  100% Acetone 30 min at room temperature

- Clearing: Clearing is carried out using propylene oxide at room temperature.
  Propylene Oxide 15 min × 2 changes
- Infiltration:
  Prop. Oxide: Embedding Medium - 3:1 (overnight)
  Prop. Oxide: Embedding Medium - 1:1 (1 h)
  Prop. Oxide: Embedding Medium - 1:3 (1 h in vacuum)
  Pure Embedding Medium 1 h at 50 °C
- Embedding: Infiltrated samples are transferred into BEEM capsules or embedding moulds and oriented. Pure embedding medium is then poured into the capsule or mould and transferred into the embedding oven set at 50 °C; kept overnight (12–24 h).

- Polymerisation: Temperature is then raised to 60 °C for 24–48 h.
- Sectioning: The sample is now cut into ultrathin sections (60–90 nm thickness) by using ultramicrotomy.
- Staining: The sample is to be stained using a double staining method (uranyl acetate and lead citrate) to give contrast to the sample in TEM.

Similarities Between TEM and SEM

For both TEM and SEM techniques, electrons are used to acquire images of samples. The main components of these two EMs are the same, i.e., an electron source and a series of electromagnetic and electrostatic lenses to control the shape and trajectory of the electron beam and electron apertures. All these components are housed inside a chamber under a high vacuum.

Differences Between TEM and SEM

TEM	SEM
TEM, as the name suggests, uses transmitted electrons, the electrons that are passing through the sample, to form a magnified image of the specimen	SEM creates an image by detecting reflected or knocked-off electrons
TEM offers valuable information on the inner structure of the sample, such as the ultrastructure of cell organelles, morphology and stress state information, crystal structure, etc.	SEM provides information on the sample's surface and its composition
TEM with the recent development, the resolution has been achieved up to less than 50 pm	SEM resolution is limited to ~0.5 nm
TEM images are 2D projections of the sample	SEMs provide a 3D image of the surface of the sample

TEM	SEM
Due to the requirement of electrons to pass through the sample, TEM samples must be fragile (generally between 60 and 90 nm ultrathin sections of the specimen) for high-resolution imaging.	For SEM imaging, there is no such specific requirement.
TEM sample preparation is quite a complex and tedious procedure that only experienced and trained technicians can follow successfully	SEM samples require little effort for sample preparation and can be directly imaged by mounting them on an aluminium stub
Acceleration voltage in TEM can be in the range of 60–300 kV	SEMs usually use acceleration voltages up to 30 kV
TEM can magnify the samples by more than 50 million times	In SEM, it is limited to 1–2 million times
TEM can only image a tiny part of the sample as only a small area (ultrathin section of the specimen) is used for viewing the sample	SEM can achieve the maximum field of view of the sample than TEMs, and the depth of field of SEM systems is much higher than in TEM systems
In TEM, the sample is located in the middle of the column. So, the transmitted electrons pass through the sample, and a series of lenses below the sample (intermediate and projector lenses) and an image is directly shown on a fluorescent screen or via a charge-coupled device (CCD) camera onto a PC screen.	In SEMs, samples are positioned at the bottom of the electron column, and electron detectors capture the back-scattered or secondary electrons. Photomultipliers are then used to convert this signal into a voltage signal, amplified to create the image on a PC screen.

Scanning Transmission Electron Microscope (STEM)

Another electron microscopy that combines TEM and SEM is called scanning transmission electron microscopy (STEM). In STEM, the electron beam is finely focused and scans the sample area as in SEM, while the transmitted electrons generate the image like in TEM. STEM can perform as both TEM and SEM systems and is an advantage of the capabilities of both techniques. Some modern TEMs can be switched to “STEM mode” and must only alter the sample alignment procedure while working with it. This instrument can look at the inner structure of the specimen with very high resolving power, higher than standard TEM resolution, but also use other signals like X-rays and electron energy loss. These signals can be used in spectroscopic techniques: energy-dispersive X-ray spectroscopy (EDX) and electron energy loss spectroscopy (EELS). Although EDX is a common practice in standard SEM and TEM systems, which is used to identify the chemical composition of samples by detecting the characteristic X-rays that are emitted from the materials when they are bombarded with electrons, EELS can only be realised in a TEM system working in STEM mode and enables the investigation of the atomic and chemical composition, the electronic properties, and the local thickness measurements of materials. These EDX and EELS are usually used in material science research.

Key Learnings

- Microscopes enhance our sense of sight by magnifying and resolving the object.
- The ability of a microscope to produce an image of an object at a larger scale than its actual size is magnification. The shortest distance between two points of a specimen that can still be distinguished as separate entities is called resolution. The resolving power of the human unaided eye is limited to 0.2 mm.
- The resolution of a microscope depends upon Abbe's diffraction limit, Airy discs, and the Rayleigh criterion.
- Condenser and objective lenses work together to create clear and detailed images of specimens in microscopy. A microscope's magnification and resolution capacity depend on these two lenses.
- LMs are simple and compound. Simple microscopes comprise a single lens, whereas a compound microscope uses numerous lenses to magnify the objects.
- In light microscopy, the illumination source is golden rays, uses glass lenses, and the magnification is about 5—2000×.
- Certain improved LMs are the darkfield microscope, phase contrast microscope, polarised light microscope, etc.

- An EM uses a focused beam of electrons instead of photons and their wave-like characteristics to magnify an object's image. The EM has variable-focus lenses, and the distance remains constant between the specimen and objective lens and the separation of the lenses. The final image produced by the EM is visualised on a fluorescent screen. It can be recorded on a photographic plate for study in a traditional EM and on a digital system in the modern EM.
- The two main types of EMs are the transmission electron microscope (TEM) and the scanning electron microscope (SEM) and another electron microscopy is a combination of TEM and SEM, called scanning transmission electron microscopy (STEM).
- In electron microscopy, the illumination source is electrons, electromagnetic lenses are used, and the magnification is about 1,500,000 (200 kV).

References

Bozzola JJ, Russel LD (1999) Electron microscopy: principles and techniques for biologists. Jones and Barlett Publishers, Sudbury, pp 48–71

Dey S, TSB B, Roy B, Dey D (1989) A new rapid method of air-drying for scanning electron microscopy using tetramethylsilane. J Microsc 156(2):259–261

Dykstra MJ (1992) Biological electron microscopy, theory, techniques, and troubleshooting. Plenum Press, New York

Hayat MA (1984) Fixation for electron microscopy. Academic Press, Inc, New York, pp 320–331

Karnovsky MJ (1965) A formaldehyde-glutaraldehyde fixative of high osmolarity for use in electron microscopy. J Cell Biol 27:137–138

Stein RS, Norris FH (1956) The x-ray diffraction, birefringence, and infrared dichroism of stretched polyethylene. Pol Sci 21(99):381–396

Microtome Device and Microtomy

3

This chapter demonstrates the types of microtomes, their uses, and microtomy techniques. The name microtome is derived from Greek micros, meaning “small”, and temnein, meaning “to cut”, a tool used to cut extremely thin slices of specimen or material called sections. These sections are used for various microscopic studies. Thereby, microtome lies in an essential position for studies in microscopy.

Learning Objectives from This Chapter

- What is microtome and its key features
- Importance of microtome in microscopy
- The invention of the microtome and its development in short
- Different types of manual, semi-automatic and automatic microtomes
- Applications of microtome

A microtome is an instrument specialised for cutting sections precisely, through which accurate and repeated slices of sections from a block of embedded tissue can be achieved. This is an essential and advanced tool for most routine histology, biological, and industrial research requiring precise tissue sectioning for

---

The original version of the chapter has been revised. A correction to this chapter can be found at https://doi.org/10.1007/978-981-97-4791-7_7

A. Haloi, *Essentials of Ultramicrotomy*,
https://doi.org/10.1007/978-981-97-4791-7_3

various applications because of its high accuracy and efficiency, ensuring exact reproducible section quality in wax and plastic-embedded blocks. Even though there are a variety of microtomes, they all have the common fundamental parts, i.e., microtome body, knife attachment, blade, and material or tissue holder. The key features in a normal microtome involve,

- Microtome base plate or stage: A platform with rails that secures the knife holder base.
- Knife holder base: This part holds the knife holder to the microtome stage. The knife holder base can be moved forward and backward from the block, which has a locked facility.
- Knife holder: This part has several components, i.e., the blade clamp that holds the blade, the knife tilt for adjusting the knife angle, and the face plate that guides those ribbons away from the blade and towards the operator.
- Cassette clamps or block holder: It holds the block in place as immobile because it moves up and down motion with each revolution. The block holder has knobs that allow the user to rotate the block face in various directions as per the need to bring the tissue in alignment with the blade.
- Coarse handwheel: Moves the block holder forward and backward to the knife.
- Advancement handwheel: Turns in one direction and advances the block towards the knife at the specified microns. Most handwheels are equipped with a safety lock to prevent the wheel from releasing and having the block holder come down towards the blade while a block is inserted or removed.
- Micron adjustment: Micron settings for section thickness can range from 1 to 60 μm on most microtomes.

Microtomy is a universal method of producing thin sections and a well-established basic biological and medical research technique. Microtomy techniques are used in microscopy in the sample preparation method, and the sections produced by this technique are used for observation under light and electron microscope. The optimum section thickness by microtomy for light microscopy depends on the size of the structural features of interest, the characteristics of the microscope being used for observation and the properties of the

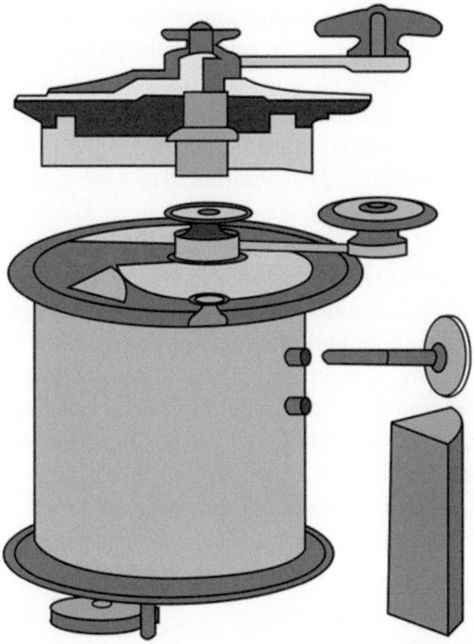

Fig. 3.1 A diagram of a microtome drawn by Cummings in 1770

polymer. Microtomy is responsible for clean, reproducible cuts through which light can be transmitted, allowing the researcher to observe the section of the sample under a microscope.

The first microtome device was invented in 1770 by George Adams, Jr. (1750–1795) and further developed by Alexander Cummings. The device was then hand operated, where the sample was held in a cylinder, and the sections were cut from the top of the sample using a hand crank. Cummings, a manufacturing company, created wooden manual sectioning instruments exclusively for cutting plants. Then, Andrew Prichard developed a table-based model to cut thin sections of wood. Although the date of the invention is uncertain, research suggests it was around 1835. This microtome was designed to cut slices of wood. Attached to a piston screw, which moves the sample up and down, the piece of wood is cut in slithers by a double-handled knife. This is the first type of microtome that does not allow any vibration during the cutting operation by affixing the device to the table with the help of a screw (Fig. 3.1).

Further development of the microtome is often credited to Wilhelm His, a Swiss-born German anatomist for the invention of the mechanical device in 1865 used to slice thin tissue sections for microscopic examination. Wilhelm His wrote in his "*BeschreibungeinesMikrotoms*" (*Description of the Microtome*) about his microtome: "*The apparatus has enabled a precision in work by which I can achieve sections that by hand I cannot possibly create. Namely, it has enabled the possibility of achieving unbroken sections of objects in the course of research*" (Fig. 3.2).

The Term "microtome" was first used by Chevalier (1839) in scientific literature. Rudolf Thoma, a pathologist from Heidelberg, and Rudolf Jung, a precision engineer, designed the first mass-produced microtome and named it "Thoma microtome", which began a new era in histology. Today, most microtomes have advanced mechanisms with a knife-block design, a changeable knife, and a specimen holder. The cutting begins by moving the sample over the knife, where the advancement mechanism allows the sample to automatically move forward so the succeeding cut

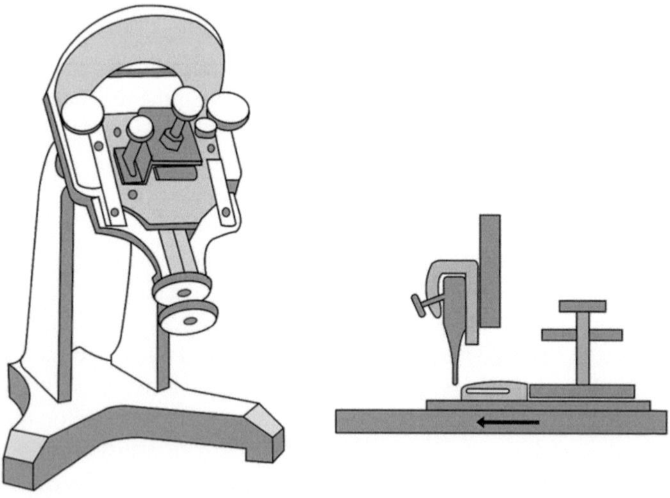

Fig. 3.2 Sketch of His microtome (1870)

for a chosen thickness can be made. Modern microtomes are designed to cut uniformly thin sections of diverse specimens for microscopic analysis, preserving their preciseness. With practice and training, fragile and translucent sections could be achieved.

Types of Microtomes

There are various types of available microtomes. The main types of microtomes are as follows:

- Manual Microtomes
- Semi-automatic Microtomes
- Automatic Microtomes

Manual Microtomes

There are different kinds of microtomes to section paraffin and plastic-embedded tissues and specialised microtomes to section frozen tissues. A sharp knife and the tissue block are held in a fixed relation in any microtome. The sections are cut between 4 and 10 μm for wax block and 0.5–3-μm plastic histological sections. The section thickness typically ranges from 8 to 15 μm for frozen sections.

Manual microtomes are rotary microtomes, rocking microtome, sledge microtome, freezing microtome, vibrating microtome, sliding microtome, saw microtome, hand microtome, etc., and have been used for traditional methods of histopathology. Manual microtomes require persistent monitoring.

Rotary Microtome

The rotary microtome is the most commonly used manual microtome. Although most microtomes are manual, some are automatic or semi-automatic, where the advancement of the block and speed of cutting is controlled by a foot pedal or a digital keypad at one's fingertips. Automatic and semi-automatic microtomes signifi-

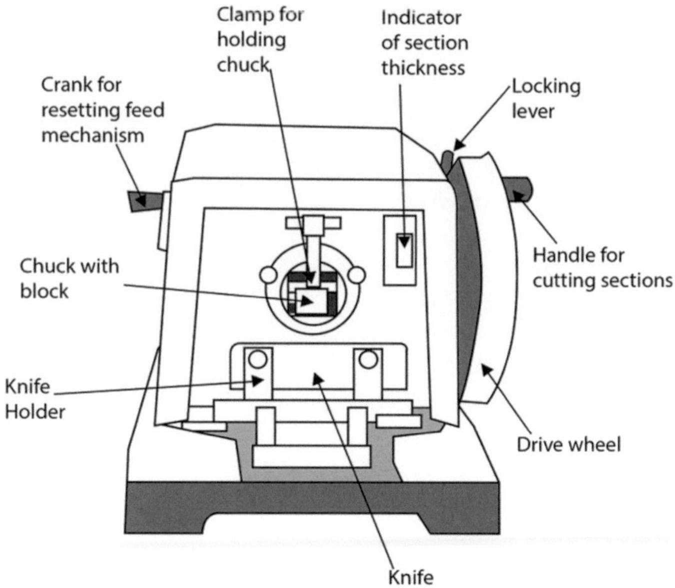

Fig. 3.3 Diagrammatic representation of rotary microtome

cantly improve ergonomics by reducing repetitive stress on joints. This device operates with a staged rotary action, so the cutting is part of the rotary motion. In a rotary microtome, the knife is typically fixed horizontally.

A rotary action of the hand wheel actuates the cutting movement. The device is heavier in weight and, thereby, more stable so that the hard tissues can be cut without vibration, and serial sections or ribbons of sections can easily be obtained. The auto-cut microtome has a built-in motor drive with foot and hand controls, and this device can produce thin sections of paraffin wax blocks about 0.5–2.0 μm. In rotary microtome, the block holder or block (depending upon the type of cassette) is mounted on the steel carriage that moves up and down and is advanced by a micrometre screw (Fig. 3.3).

Advantages

- The machine is heavy, stable, and does not vibrate during cutting.
- Serial sections can be obtained.
- Cutting angle and knife angle can be adjusted.
- It may also be used for cutting celloidin-embedded sections with the help of a unique holder to set the knife.

The above figure explains the principle of cutting sections with a rotary microtome. Through the motion of the sample holder, the sample is cut by the knife, and the fresh section remains on the knife. At the highest point of the rotary motion, the sample holder is advanced by the same thickness as the section, allowing for the next section to be made. The typical cut thickness for a rotary microtome is between 1 and 60 μm. This microtome design can allow for reasonable "semi-thin" sections with a thickness of as low as 0.5 μm for hard materials, such as a sample embedded in a synthetic resin.

Sledge Microtome

Sledge microtome is used for preparing large samples, such as biological specimens embedded in paraffin and resin-embedded material, including whole organs, for light microscopy. The section thickness is generally between 10 and 60 μm. In this category of microtome, the sample is placed into a fixed holder (shuttle), and the sledge is placed upon a linear bearing. This design allows the microtome to cut many coarse sections readily. The sledge microtome has the advantage of cutting large and complex blocks because of its knife-holding clamps, which will enable the knife to be offset to the direction of the cut. It is cumbersome for stability and not usually subject to vibration. Besides biological applications, sledge microtome can also cut material samples from various industrial applications like wood, plastics, textile fibres, etc. (Fig. 3.4).

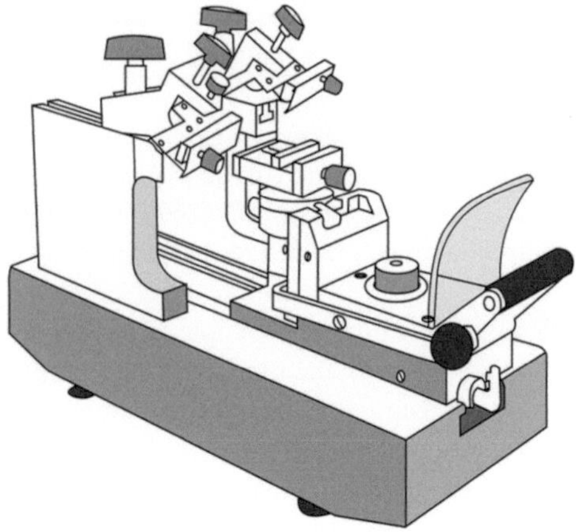

Fig. 3.4 Diagrammatic representation of sledge Microtome

Cryomicrotome and Freezing Microtome

A cryomicrotome is a sectioning device of organic tissues for microscopic examination at very low temperatures. Rotary microtomes can also be adapted to a liquid nitrogen chamber and a cryomicrotome setup to cut frozen samples. The low temperature increases the hardness of the sample by undergoing a glass transition to prepare semi-thin sections. In this sectioning technique, the sample and the knife's temperature must be controlled to optimise the resultant sample thickness, i.e., up to 20 μm of frozen tissue. A cryomicrotome possesses five parts: a freezing shelf, specimen holders, microtome, blade holder, and anti-roll guides. Liquid helium or nitrogen is used to achieve low temperatures. It is a highly precise instrument that maintains the cryogenic temperature of samples or devices placed inside it and is also known as a Dewar flask. Cryomicrotome is useful in medicine, science, and engineering for microscopic examination to study the margin of cancers, quick diagnosis of tissue sections, examine enzyme histochemistry to diagnose and treat neuromuscular diseases, histopathology, and immunohistology.

Freezing microtome is used with high precision to cut thin to semi-thin sections of fresh, frozen tissues of industrial products such as textiles, paper, leather, soft plastics, rubber, powders, pastes, food products, etc., with high precision and CO_2 freezing attachment. It has a stage where tissues can be quickly fixed and frozen using either liquid carbon dioxide from a cylinder or a low-temperature re-circulating coolant. Through this process, the water-rich tissues are hardened and are sectioned in that frozen state. Further, these sections are stained and observed under a light microscope. This procedure is applied for quick diagnosis in medical science in immunohistochemistry (IHC). Freezing microtomes is useful for biological analysis as tissue does not alter or mask the chemical composition.

Cryomicrotome and freezing microtome, although process frozen tissue samples, a freezing microtome is used to make thin sections of frozen tissues, and a cryomicrotome maintains the cryogenic temperature of samples or devices that are placed inside it for microscopic studies (Fig. 3.5).

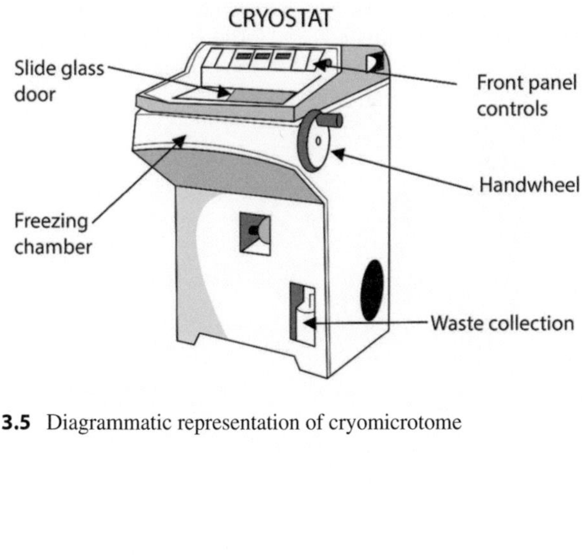

Fig. 3.5 Diagrammatic representation of cryomicrotome

Vibrating Microtome

In a vibrating microtome, biological tissue cuts accurately and efficiently without the requirement of freezing or embedding and still maintains the tissue's cell morphology, enzyme activity, and cell viability. This microtome stands in an advantageous position because sections can be achieved without any additional processing, fixing, or freezing from fresh biological tissue. In contrast, generally, in other microtomes, tissue must be fixed chemically and embedded in paraffin, wax, or epoxy resin and frozen using the cooling chamber before sectioning. Vibrating microtomes are one of the most preferable and used tools in enzyme histochemistry and rodent neuroscientific research because they can sustain the tissue's viability, enzyme activity, and cell structure. The vibrating microtome is usually used for different biological samples, and it operates using a vibrating blade, allowing the resultant cut to be made with less pressure than would be required for a stationary blade with a thickness usually around 30–500 μm for live tissue and 10–500 μm for fixed tissue (Fig. 3.6).

Saw Microtome

Saw microtomes are designed to cut hard and brittle materials such as teeth, bones, minerals, or artificial biomaterial devices.

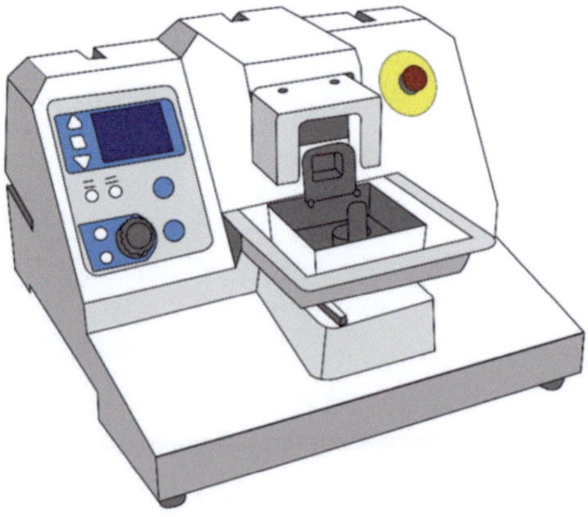

Fig. 3.6 Diagrammatic representation of vibrating microtome

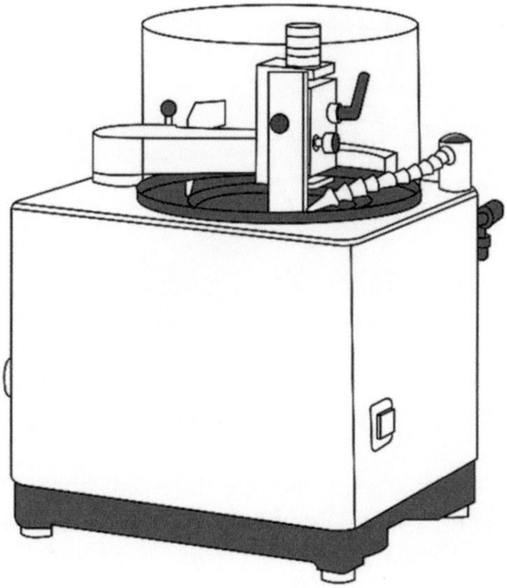

Fig. 3.7 Diagrammatic representation of saw microtome

This microtome has a recessed rotating saw, which slices through the sample. They cut through the samples by pushing the sample across a recessed rotating saw, and the section thickness is approximately 30 μm and can be made for comparatively large samples. This device is used to prepare histological sections from non-decalcified bone tissue, even in the presence of various metal, ceramic, and polymeric medical devices. It is helpful in research and testing labs in Archaeology, Histology, Geology, Forensics, Pathology, Implantology, Orthopaedics, and Dentistry, as well as in the medical device industry (Fig. 3.7).

Sliding Microtome

The specimen block is fixed, and the knife is carried obliquely across it. It is sturdy equipment for sectioning paraffin, cellodine, embedded specimens, or in natural form. This microtome can cut sections of relatively large pieces of tissue about 40–50 μm thick. The sliding microtome is designed so that it causes few breakdowns and allows for the regulation of the blade's pressure on the

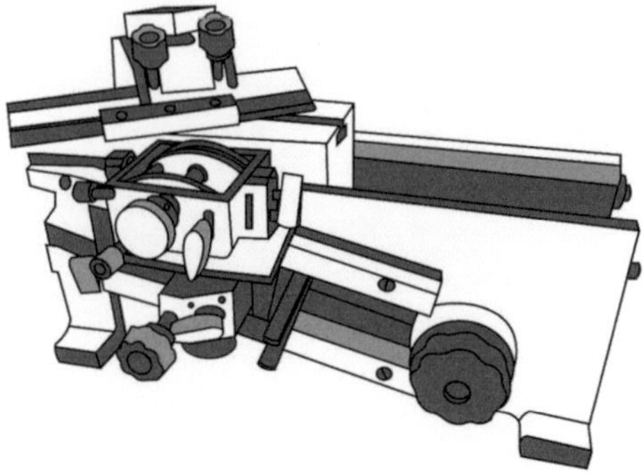

Fig. 3.8 Diagrammatic representation of sliding microtome

specimen block. Because of the size and arrangement of the blade, it is possible to section large-head blocks and cut blocks of celloidin. The sliding microtome is a hands-on tool for scientific studies. It is used principally for sectioning cellulose nitrate-embedded tissue, while rotary, rocking and sledge microtomes are widely utilised for paraffin wax-embedded material. Sliding microtome is the excellent and most popular microtome generally used for cutting serial sections where the thickness can be adjusted according to the need (Fig. 3.8).

Hand Microtome

A hand microtome is a basic form used to cut thin sections, such as specimen parts of plants and animals, by hand quickly and accurately. This microtome is suitable for sectioning of fresh tissues. The hand microtome comprises a knife holder base, knife holder, microtome body, cassette clamp or block holder, advancement hand wheel, coarse hand wheel, micron adjustment knob, safety lock, and a single-edged razor blade. Hand microtome can be securely attached to the edge of a table bench or any work surface with the help of a fixing clamp. Sections are usually cut in steps of 10 μm each. This microtome is used for fieldwork during biology excursions and practical microscopy classes. The disadvantage of this microtome is that it cannot produce serial sections, so the sectioning process slows down (Fig. 3.9).

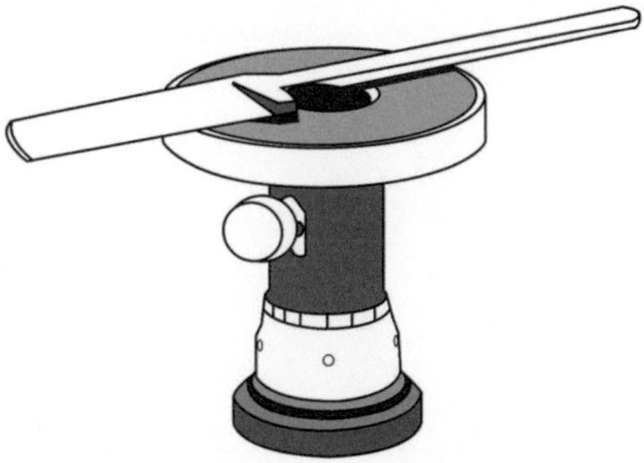

Fig. 3.9 Diagrammatic representation of hand microtome

Semi-Automatic Microtomes

Semi-automatic microtomes are ultramicrotomes and cryo-ultramicrotome.

Ultramicrotome (UM)

UM is an indispensable tool for the preparation of extremely thin sections, essential for TEM and Serial Block-Face Scanning Electron Microscopy (SBFSEM), and also sometimes for light-optical microscopy. Because of the mechanical construction and varied tight tolerances of ultramicrotome, the linear thermal expansion of the mounting is used to provide very fine control of the section thickness. UM can produce UT sections of 60–90 nm thickness, ideal for transmission electron microscopy and often between 30 and 50 nm for SBFSEM. Semi-thin (ST) sections of about 400–500 nm are also taken for light microscopy survey sections to select an area for the final UT sections. In ultramicrotomy, a ribbon of UT sections is prepared at room temperature and floats on water in the boat of a diamond knife used to cut the sections. The sections are then collected onto a Cu grid for TEM study. The main advantage of UM is that it produces UT sections of homogeneous size of the electron-transparent area within the sections in the succeeding sections. Ultramicrotomy can be used for biological (animal and plant) and material samples (rubber and plastics,

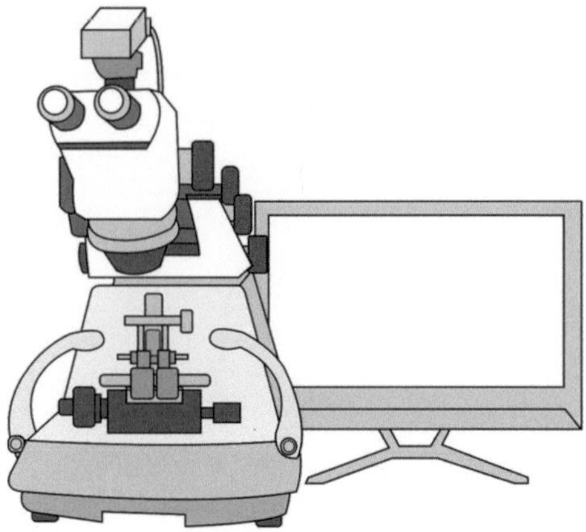

Fig. 3.10 Diagrammatic representation of ultramicrotome

etc.) (Fig. 3.10). Ultramicrotomy techniques will be briefly described in the next chapter.

Cryo-ultramicrotome

A cryo-ultramicrotome is similar to an ultramicrotome and is used to prepare semi- and ultrathin sections of biological samples that can be examined under the electron microscope attached to a cryo-setup. In cryo-ultramicrotome, a cryo-chamber is available to cut biological samples, and the technique is carried out at freezing temperatures between 15 °C and −185 °C, where the section thickness varies from 50 to 200 nm depending upon the need. This technique is used to cut ultrathin frozen biological specimens. The main advantage of cryo-ultramicrotome is speed, as the sample preparation of a specimen is faster than the resin-embedded method in ultramicrotomy. The section thickness achieved practically is about 100 nm, thus limiting the ultrastructural studies (Fig. 3.11).

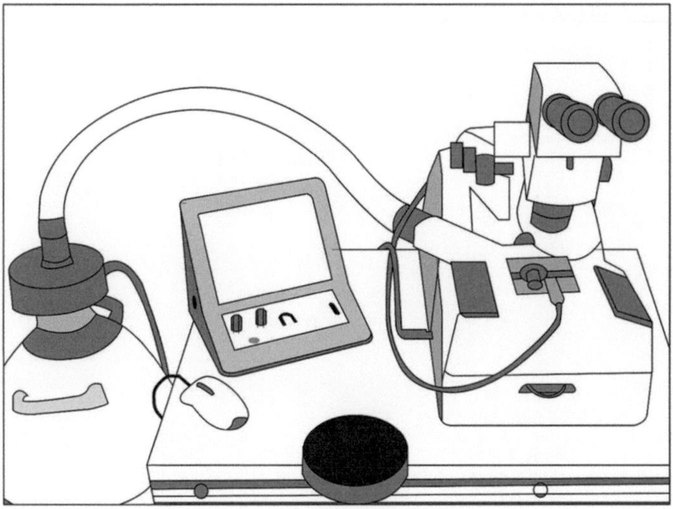

Fig. 3.11 Diagrammatic representation of cryo-ultramicrotome

Automatic Microtomes

Automatic microtomes are laser microtomes, and computer microtomes used to save time.

Laser Microtome

The laser microtome is an instrument for contact-free sectioning used for rigid materials, such as bones or teeth, and some ceramics. The section thickness is between 10 and 100 µm dependent upon the properties of the sample material. In laser microtome, the exchange of tools for each separate cut is not required, and the same setup can produce different shapes for the same specimen with high precision. It is much quicker than traditional mechanical sectioning methods as prior sample preparation through embedding, freezing, or chemical fixation is not required, thereby minimising the artefacts from preparation methods. The device operates using the cutting action of an infra-red laser, and the combination of high power with a high raster rate allows the scanner to cut large sample areas quickly. This device is often costly,

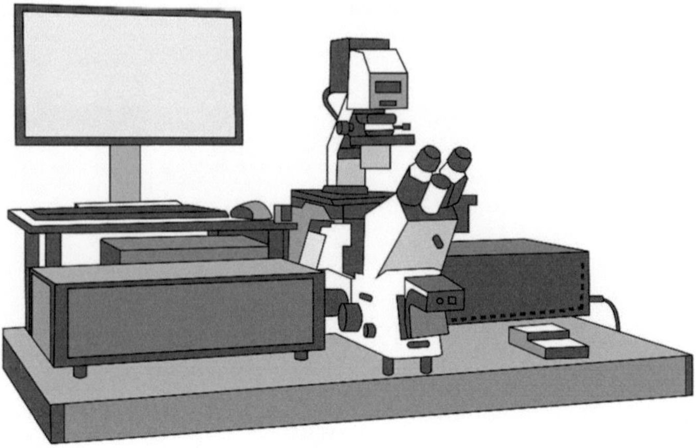

Fig. 3.12 Diagrammatic representation of laser microtome

and a proper ventilation system is a requirement for a safe working environment (Fig. 3.12).

Computer Microtome

The computer microtome is an instrument used in biopsy histopathology. Due to its sophisticated electronic and mechanical design, the speed of the automatic slice of tissue sections can be adjusted, and it can also switch manual and electric smart chips quickly. This instrument has a security emergency braking system. Its advanced actuation system can slice the tissues more precisely, reliably, and quieter, and the LCD enables one to see the slice and trimming thickness. The hand wheel can be locked at any position according to need. It also has a safety alarm system, and the waste bath can be easily disassembled. The sections are cut at about 0.5–100 μm (Fig. 3.13).

Applications of Microtome: The most common applications of microtome are,

Traditional Histology Techniques

Histology is essential to evaluate the morphology and structure of cells, tissues, and organs under the microscope. Through histol-

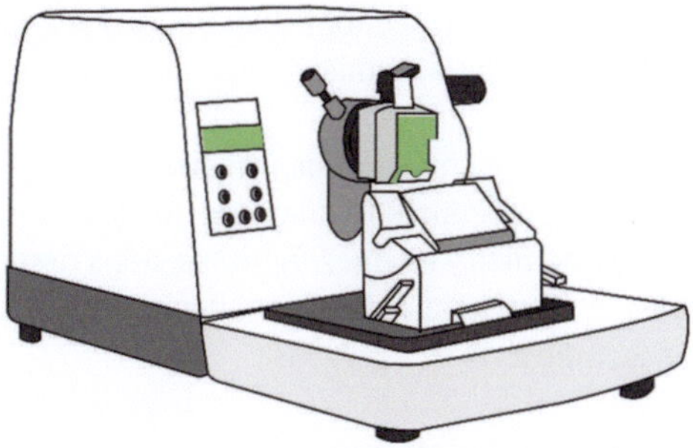

Fig. 3.13 Diagrammatic representation of computer microtome

ogy, it has become possible to observe tissues and cells for the first time and distinguish between their single components, thereby detecting disease based on cell morphological alterations or observing pathogenic agents like bacteria in tissue samples. Microtome is an essential part of sample preparation in histology, as thin sections are needed for microscopic studies. Before tissue sectioning in a microtome, sample processing by fixation, dehydration, clearing, wax infiltration, and embedding is done. The tissue is then cut in the microtome at a thickness varying from 2 to 50 μm. From there, the tissues can be mounted on a microscope slide, stained with appropriate aqueous dye(s) after prior paraffin removal, and examined using a light microscope for histological study.

Cryosectioning Technique

The cryosectioning technique comparatively invests less labour and time than the histological technique and provides a good system for visualising fine details of the cell. Sample preparation of cryosections does not involve the dehydration step, and labelling, and observation of specimens can usually be carried out in 1 day. In general, the sample is frozen either in isopentane or liquid nitrogen. Rapid freezing reduces ice crystal formation and minimises morphological damage. After the sample hardens by freezing, it is cut into the frozen state with a freezing microtome or cryomicrotome. The sections are then stained and examined with a light microscope. Frozen sections may be used for various procedures, including immunochemistry, enzymatic detection, and in situ hybridisation.

This technique is much faster than traditional histology and is used with medical procedures to diagnose quickly. It is also used in immune histochemistry as freezing tissue stops tissue degradation faster than a fixative and does not alter or mask its chemical composition as much.

Electron Microscopy Technique

As with the traditional histology technique, after embedding tissue in epoxy resin, a microtome equipped with a glass or gem-grade diamond knife cuts very thin sections, typically between 60 and 90 nm, called UT sections. UT sections are cut with UM, suitable to view under TEM with a resolution of 0.1 nm using 200 kV TEM. Sections are then stained with an aqueous solution of an appropriate heavy metal salt, and the ultrastructure of the sample is viewed/observed with a TEM. The UM is also used to cut ST sections for histology slide preparation and survey sections before UT sectioning with its glass knife or an industrial-grade diamond knife. The survey sections are 400–500 nm thick, mounted on a glass slide, and stained to locate areas of interest under an LM before UT sectioning for the TEM.

Key Learnings

- *A microtome* is a specialised instrument that can produce accurate and repeated slices of sections precisely from a block of embedded tissue.
- The critical features in a normal microtome involve a microtome base plate or stage, knife holder base, knife holder, cas-

sette clamps or block holder, coarse handwheel, advancement handwheel, and micron adjustment.

- Microtomy is part of sample preparation for microscopy and a well-established basic technique adopted as a universal method of producing thin sections for biological and medical research.
- George Adams, Jr. invented the first microtome device in 1770, which Alexander Cummings further developed. This device was then hand-operated.
- The three main types of microtomes are manual, semi-automatic, and automatic microtomes.
- Manual microtomes are rotary microtomes, rocking microtome, sledge microtome, freezing microtome, vibrating microtome, sliding microtome, saw microtome, hand microtome, etc.
- Semi-automatic microtomes are ultramicrotomes and cryo-ultramicrotome.
- Automatic microtomes are laser microtome and computer microtome.
- The most common applications of microtome include traditional histology techniques, cryosectioning techniques, and electron microscopy techniques.

4 Ultramicrotome Device and Ultramicrotomy

In the previous chapter, we covered the invention of microtome and its type and applications. As time advances, the microtome also modified and improved, thereby introducing an ultramicrotome where ultrathin (UT) sections are cut to study the ultrastructure of cellular organelles with its fine details in a transmission electron microscope (TEM) or scanning transmission electron microscope (STEM). UT sections are very thin slices of the specimen, sufficient to transmit the electron beam. Good quality UT sections (usually between 60- and 90-nm silver coloured) can provide a good image of the sample in TEM. It is imperative to cut UT sections of specimens to achieve images of the ultrastructure of biological samples or some specific material sample to conduct a TEM study. The method of cutting specimens into UT sections is called ultramicrotomy and is applied for biological sample preparation, both animal and plant tissue and some industrial materials such as polymers, membranes, plastic, and soft metals, etc., used for visualising and analysing the internal fine structure of samples/specimens at nanometre scale resolution, that can be studied and documented at different magnifications in a TEM or STEM. Ultramicrotomy procedure lies in a very crucial but essential position for TEM or STEM study. Only a perfect section can produce a perfect micrograph of a specimen in TEM or STEM, and a perfect block can reach a perfect section if all the parameters/ steps of ultramicrotomy procedures are performed flawlessly, which are relatively critical and tedious.

A. Haloi, *Essentials of Ultramicrotomy*,
https://doi.org/10.1007/978-981-97-4791-7_4

Learning Objectives from This Chapter

- What is ultramicrotome (UM) and ultramicrotomy
- History of UM
- Principle of UM
- Application of UM
- Advantages and disadvantages of UM

The ingenious idea of discovering microtome and then UM to cut UT sections on a nanometre scale though looks easy to attend to and witness working, but overcoming the finer details to achieve successfully operating UT sections can take longer time and sometimes years. UT sectioning of tissues or specimens is an art. For sectioning, the selection of knife material (whether glass knife or diamond knife), blade shape, cutting speed, knife angle, and other variables varies for different types of specimens, and it is determined through experience. Sections cut under non-optimal conditions show tearing, ripping, rupture, hole, folding, etc.

The ultramicrotome is a mechanical instrument used for cutting UT sections. *Ultramicrotomy* is the procedure through which a specimen (biological/material) can be cut into extremely thin slices, which permits the electron beam to pass through in TEM or STEM and so can achieve the internal fine structure of the sample/specimen, which can be viewed and analysed at nanometre scale resolution and also documented at different magnification in TEM or STEM. The electrons of standard TEM of 50–200 kV cannot pass through biological specimens much thicker than 150 nm, and an ultramicrotome can cut ultrathin sections of 5 nm onwards. For best resolution in TEM, UT sections are to be cut from 60 to 90 nm thick, roughly the equivalent of cutting a single red blood cell into 100 slices. This technique can achieve ST (usually 400–500 nm) and UT sections.

History of Ultramicrotome

The development of ultramicrotome constituted the significant technical advance in TEM and STEM. This advance required not only the construction of microtomes capable of producing sections thin enough to be viewed in a TEM or STEM but simultaneously the development of improved chemical fixation, dehydration, clearing and discovery of embedding resins, and staining protocols and selection of specially sharpened glass knife to Diamond knife. Some of the known history and gradual development of ultramicrotome are tabulated below:

1939	Von Ardenne	Modified a histological rotary microtome to cut wedges, the edge of which was thin enough to allow the electron beam to penetrate them.
Early 1950s	Porter and Blum in the USA	Single-pass ultramicrotomes capable of cutting sections of uniform thickness were developed. This ultramicrotome has mechanical advances and is manually operated, with only a single fluorescent overhead light by RMC.
1950s	Sitte in Austria	Developed the Reichert OM-01. This instrument was thermally advanced and cut sections automatically.
1957	British Huxley MK 1	It was a manual instrument with a mechanical advance that was marketed.
Early 1960s	Swedish company LKB	They marketed the ultramicrotome-I, which has thermal advances in an automatic machine with a magnetically retracted arm that allows single-pass cutting.
1962	Porter and Blum	They developed the mechanically advanced, automatic MT-2, which quickly developed a reputation for reliability and stability.
1964	Reichert OM-U2	This unit was a thermally advanced automatic instrument with an adjustable cutting window and variable cutting stroke and return speeds.

1965	Ultrotome III by LKB	This instrument had a novel lighting arrangement, incorporating both an incandescent and fluorescent light in an overhead boom that could be swung over a wide area from behind the knife to in front of the knife, depending on whether the user wanted to see the interference colours of the sections or a good reflection in the black face during the approach of the knife to the specimen.
1970	Mk-2 by Huxley	It was a mechanically advanced, automatic machine
1970s	OM-U3 by Reichert	It had a backlight for specimen advance and various minor modifications in the knife stage controls and specimen adjustment controls.
1970s	MT-2B from DuPont/Sorvall	It had an adjustable return speed for the first time.
1970s	Ultrome IV by LKB	This instrument was produced in NOVA, which abandoned the vacuum tube technology of the Ultrome III for more stable solid-state technology.
Late 1970s	New generation ultramicrotome or cryo-ultramicrotome	It had more excellent mechanical stability to meet the increasing interest.
Late 1970s	MT 5000 by Dupont/Sorvall	It had an adjustable cutting window for the first time, and a stepping motor-controlled mechanical advance for the automatic cutting cycle.
1978	Ultracut (OM-U4) by Reichert	This instrument had improved to mechanical advance from their traditional thermal advance. Sitte had developed this unit.
1980s	LKB Nova, Reichert Ultracut-E and DuPont–Sorvall MT-6000	The present generation of ultramicrotomes
From 1985 onwards, a series of corporate merged, resulting in the ultramicrotome of MT-7000 from RMC.		

Principle of Ultramicrotome

To achieve an image of a specimen in TEM, an electron beam has to pass through the specimen, and the permeability of the electron beam through the specimen depends partially on the specimen mass and thickness (thickness × density) and also on the acceleration voltage of the EM. Usually, a biological specimen much thicker than 150 nm cannot penetrate electrons of standard TEM. So, the need for UT sections of the specimen arises, and UM can cut UT sections of 5 nm onwards. Sections cut between 400 and 500 nm are used as ST, which is used for optical microscopic study, and sections between 60 and 90 nm are ideal for TEM observations.

Before preparing to cut UT sections, the work of an ultramicrotome device and the function and the knob of the control unit are required to know. UM comprises two main parts: the ultramicrotome device (sectioning unit) and its control unit (Fig. 4.1).

The UM device or sectioning unit comprises many elements. Starting with the hand wheel, which causes the up and down

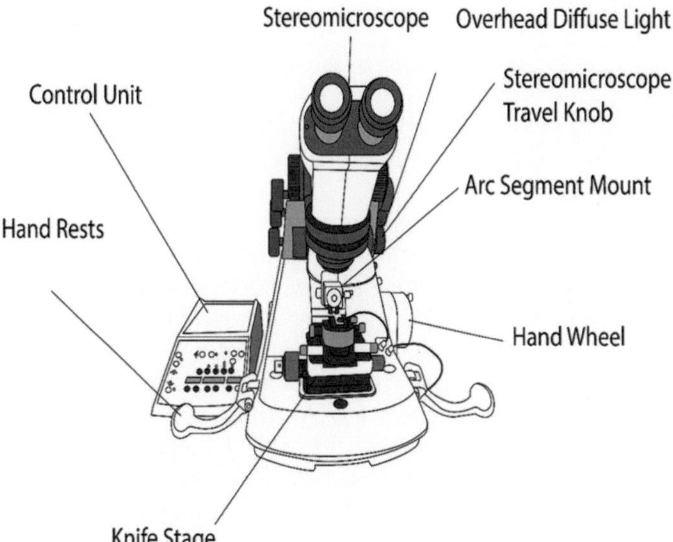

Fig. 4.1 Sketch of ultramicrotome

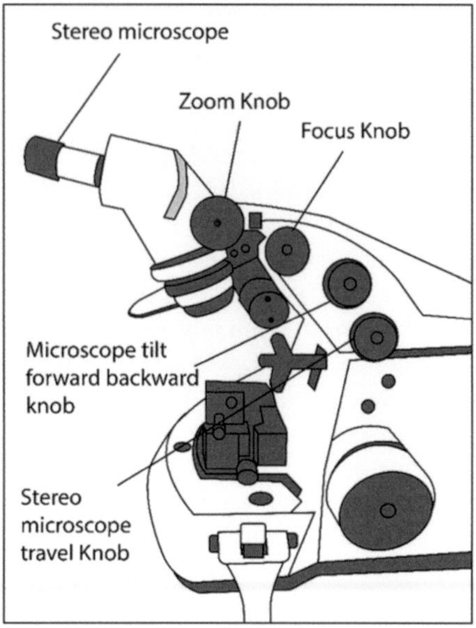

Fig. 4.2 Diagrammatic representation of stereomicroscope adjustments of ultramicrotome

moves of the cutting arm to go through a complete stroke where the upstroke corresponds to the retract move by a minimum of 80 μm of the cutting stroke. Likewise, the cutting arm extends outward to cut a section from the specimen block in the down move (Fig. 4.2).

Furthermore, a stereomicroscope and holder are also present with many adjustment facilities to achieve and retain an optimal view of the specimen cutting area, which helps observe and calibrate the device during the sectioning. The stereo microscope can move to the left and right and tilt up and down, allowing the microtomist to set the optimum view angle or zoom in and out according to the requirements. A stereo microscope is also attached with a large overhead diffuse light, illuminating the cutting area brightly. Another LED lighting system is present in the upper stage for backlighting or transilluminating the knife edge and specimen during knife adjustment (Figs. 4.3 and 4.4).

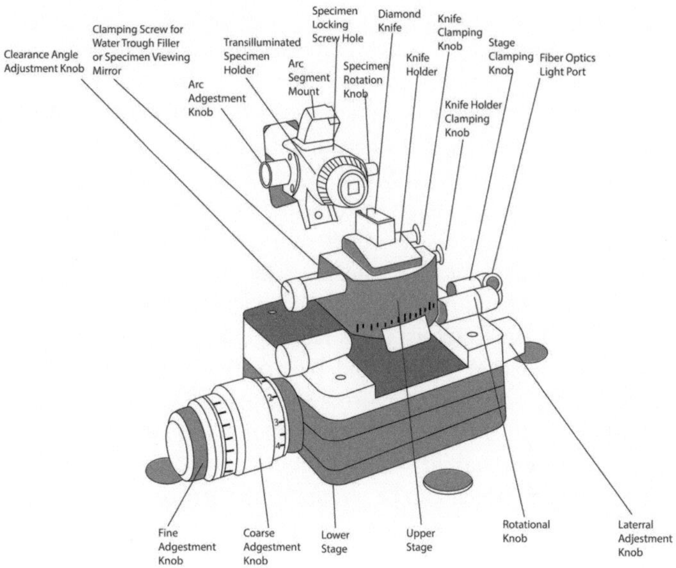

Fig. 4.3 Diagrammatic representation of upper and lower stage and arc segment holder of ultramicrotome

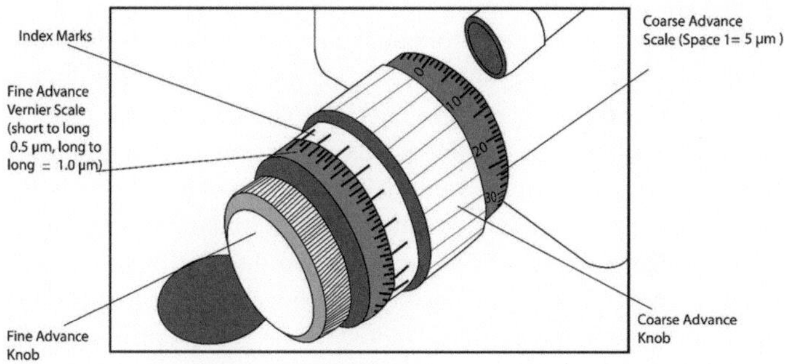

Fig. 4.4 Diagrammatic representation of fine and coarse micrometer advance knobs of ultramicrotome

A knife stage is positioned in the sectioning unit, and it is of two parts, upper stage, which is located on the top of the lower stage, is clamped firmly into position by a clamping knob. By releasing this knob, the upper stage can slide forward and backwards. This feature is used when aligning the position of the knife edge and specimen face together, which is 1 or 2 mm apart from the knife and the specimen, before starting the final approach of the cutting procedure. The upper stage, which contains a knife holder where the knife is mounted, is facilitated with 360° rotation while attached to the lower stage, providing many movement options to the microtomist. This feature is mainly used for trimming and preparing a trapezoid face of the block, which later helps achieve serial sections from the specimen block and view the cutting knife's back side. Consequently, the lower stage provides maximum stability and incorporates a system of special roller bearings which can be shifted to the left or right with the lateral knob present. The lower stage bears the coarse and fine advanced knobs, which have graduation marks, measured in μm or microns can precisely forward and backwards the upper stage containing a knife to the specimen. The microtome device has an arc segment mount to fit in the specimen holder. Additionally, UM offers plenty of angle movements, which are used to cut high-quality sections (Fig. 4.5).

The control unit or control panel consists of many buttons that regulate the sectioning unit to cut specimens of different dimensions, hardness, and types of materials. The control unit is separated into many parts, which are used for specific operations of the UM. All the buttons have their unique operation; one is the motorised cut button, which controls the automatic cutting mode of UM. There are two LED lamps above the Cut button, which display and indicate the position of the cutting arm, whether inside or outside the cutting window. If the green lamp is lit, the cutting arm is in the cutting window, and the cutting arm is outside if the yellow lamp is lit. The dual foot pedal also serves as a cut button from the control panel. A cutting window is an area or stretch where the section from the specimen will be cut, and the microtomist can maintain the cutting speed and thickness of the section according to the need. The cutting window can be set by

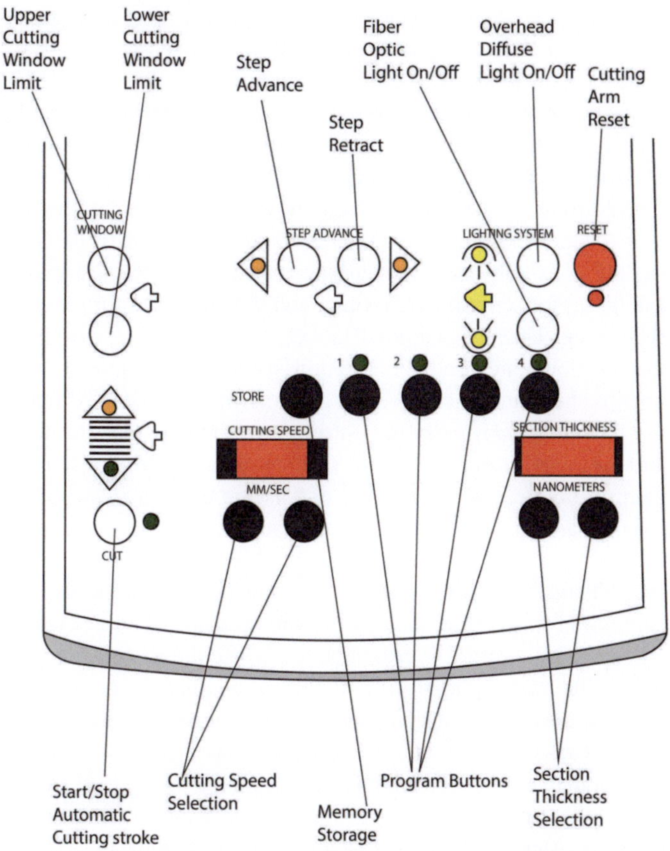

Fig. 4.5 Diagrammatic representation of control panel of ultramicrotome

depressing the start and end buttons from the control panel. When the instrument is in automatic cutting mode, depressing these buttons will not work. To use these buttons, the automatic cutting mode must be off. A button to set the cutting speed in mm/s can be changed by depressing + or – to maximise or minimise, respectively. Specimens embedded in epoxy resin normally have the best results when seen at speeds between 0.5 and 1.5 mm/s. Cutting speed is effective when the specimen arm passes through the cutting window, also called a cutting zone. After a section is cut through the cutting zone, the cutting arm retracts at the set retract speed. Section thickness in nanometres (nm) is indicated in the red display, and there are five buttons where the value can be stored for section thickness along with its cutting speed on the control unit, which is called the memory storage button.

Additionally, the thickness may be changed by depressing the + or – button and 1-nm incremental changes are made by depressing either button once. For cutting convenience, the last button can be set for semithin sectioning, and the others are for ultrathin sectioning. Usually, diamond knives, which produce extremely thin sections, are used for ultrathin sectioning because of their sensitiveness and expansiveness. Furthermore, in step advance, the left button advances the cutting arm towards the specimen, and the right button moves away the cutting arm from the specimen; in both cases, the amount of forward and backward of the cutting arm is displayed in the section thickness display. Continuing, a reset button is present, which will reset the cutting arm advance mechanism to the starting point of its travel. By depressing this button, the red LED lamp below the reset button will remain lit until it is complete to reverse the advance mechanism. The advance totaliser and section counter button displays the total distance in nm. The arm has advanced since the display was last reset. The on and off buttons of the lighting system are present in the control panel of the UM. Among the optional accessories, there is a specimen viewing mirror used for viewing the block face during trimming, and it is positioned on the upper stage by inserting its base peg into the accessory hole on the top left side of the upper stage. The water trough filler also stands as an optional accessory used to fill the troughs of glass and diamond knives with distilled water very precisely.

An UM is designed to cut UT sections of specimens mechanically. The specimen arm of UM performs a reciprocating movement, and steps forward the distance that is introduced for feeding when the sample is in its higher position, and it is done either manually by turning the hand wheel clockwise with a constant low speed or in automatic mode. The cutting starts when the sample is just above the knife edge. After sectioning, the arm moves the sample below the knife edge at the bottom point and returns at the return speed along with 1 mm backward movement to cross the knife edge and perform sectioning again.

In ultramicrotomy, selecting a good knife for specimen sectioning is very important. A glass knife or a diamond knife is used to achieve semithin and ultrathin sections in ultramicrotomy. Latta and Hartmann (1950) discovered that the edge of broken glass or glass knives could be used to cut sections from specimens in ultramicrotomy. Glass knives then became very popular in ultramicrotomy for trimming block faces and also cutting semithin and ultrathin sections of specimens. In 1953, Humberto Fernandez-Moran discovered that a gem-quality diamond could be used as a more durable knife edge for cutting ultrathin sections in ultramicrotomy, as the diamond was the hardest known material then. A diamond knife edge is more durable than a glass knife. The edge of a glass knife dulls too quickly, and a small number of ultrathin sections can be cut with a glass knife, whereas a diamond knife can be used to cut thousands of sections on one area of the knife.

Before the sectioning process begins, the polymerised block (capsulated or flat block) has to be hand-trimmed with the help of a block trimmer and a razor blade to remove the excess resin surrounding the specimen. Then, the specimen block must be mounted on the block holder of the Arc segment mount of UM and then be tightened by the specimen locking screw of the arc segment. Fine trimming is done in UM with glass knives. For that, the knife should be moved forward to approach the specimen block very gently, and the face of the specimen block is usually trimmed to a square, rectangle, or trapezoid shape. The trapezoid shape of the block face is needed when the ST sections are prepared before UT sectioning. After trimming, the glass knife with trough should be mounted on the knife holder, and the trough should be with doubled distilled water using a Reflexomatre filler syringe or by water trough filler. Through the stereomicroscope of the UM, the microtomists can focus on the knife edge and inflate or deflate water from the trough to create a mirror-silvery surface. A single drop of filtered Polysorbate-20 solution can also be added to the water to lower the surface tension. This helps to keep always the knife edge wet. Also, due to the evaporation of water, the knife trough must be frequently filled with water so that the surface is silvery at all times when cutting. It is necessary to keep the water level stable while cutting to see proper interference of colour of the sections, due to which section thickness can be maintained according to the need. The interference colour becomes silvery when the thickness is approximately 60 nm, as the peak is almost out of the visual light range. As the sections' thickness minimises, the interference colour becomes grey as the peak moves out of the optical range. As the knife moves closer to the block face, the knife edge reflection becomes thinner and retreats. On the other hand, if the line is not parallel to the knife edge, specimen rotational and tilt alignment must be done. The sectioning is performed by a vertical movement of the specimen over the extremely sharp blade of a fixed glass or diamond knife. Removing the sections directly from the knife blade is difficult because they are so thin. So, they are allowed to float on the water surface of the trough and are collected from the surface of the water bath after the sectioning procedure. Semithin sections (approx. 400–500 nm) are cut first to examine under a light microscope and locate the area of interest. Then, keep only the area of interest to be viewed under TEM, and the rest is trimmed off again following the same trimming procedure. Cut ultrathin sections about 60–90 nm thickness. ST sections are picked up onto a glass slide and stained with toluidine blue, and UT sections are picked up onto grids and stained with uranyl acetate and lead citrate before observing under TEM.

Applications of Ultramicrotome

The main application of ultramicrotome is to produce UT sections for transmission electron microscopy (TEM) and scanning transmission electron microscopy (STEM) and also ST sections for light microscopy. The samples for ultramicrotomy include biological specimens (animal and plant tissues) and industrial materials (polymers, membranes, etc.). UT sections are indispensable for TEM examination as samples must be thin enough to penetrate the electron beam through the sample. UM sectioning reduces the specimen thickness to the nanometre scale. So, the specimen section qualifies for observation in TEM or STEM to view the internal fine structure at the cellular level of the specimen in the case of the biological sample, and any material sample sections yield atomic resolution from the core region of particles of hundreds of nanometres. Ultramicrotomy is a mechanical cross-sectioning method that introduces no observable microstructural artefacts or structural damage, like microcracking and nanoporosity. Both ST and UT sections can be achieved by the ultramicrotomy technique. A quality UT section results in a high-quality TEM image, thereby providing statistically significant results of the TEM analysis.

This method is a valuable model for many diverse scientific, educational, and industrial fields, such as life sciences, nanotechnology, medical, biological, and material research, forensic analysis, gemology, and metallurgy. It facilitates studying topographical, morphological, compositional, and crystalline information of samples in TEM. UT sectioning makes it possible to analyse the ultrastructure and texture of samples by making the study available on a molecular level. Companies also use this technology to identify flaws, fractures, and damages to micro-sized objects, and the data attained from this research can help fix problems or make a more durable, efficient product.

Advantages

As UM can produce UT sections, it offers a very high magnification of samples in TEM, which can be utilised in various scientific, educational, and industrial fields. High-quality UT sectioning by UM helps in acquiring high-quality TEM micrographs, so it becomes possible to achieve detailed information on the element and compound structure of the material, which can yield insight into the surface features, shape, size, and structure sample as well as ultrastructure of a biological sample.

Disadvantages

UM is an expensive and heavy instrument and requires a vibration-free environment to work with. Furthermore, the ultramicrotomy procedure involves laborious sample preparation prior to the final sectioning of samples, and the whole operation procedure needs special training. It also requires constant maintenance.

Key Learnings

- *Ultramicrotome* is a mechanical instrument used for cutting UT sections, and *ultramicrotomy* is the procedure through which a specimen (biological/material) can be cut into extremely thin slices. UT sections permit the electron beam to pass through it to achieve the internal fine structure of the sample/specimen, which can be viewed and analysed at nanometre scale resolution and documented at different magnifications in TEM or STEM.
- The gradual development of UM not only constituted the construction of microtomes capable of producing sections thin enough to be viewed in a TEM and STEM but simultaneously the development of improved chemical fixation, dehydration, clearing, and discovery of embedding resins and staining protocols and selection of specially sharpened glass knife to Diamond knife.
- Usually, a biological specimen much thicker than 150 nm cannot penetrate electrons of standard TEM. So, there arises the need for UT sections of the specimen, and UM can cut UT sections of 5 nm onwards. Sections cut between 400 and 500 nm are used as ST, which is used for optical microscopic study, and sections between 60 and 90 nm are ideal for TEM observations.
- The main application of UM is to produce UT sections for TEM and STEM and ST sections for LM. The samples for ultramicrotomy include biological specimens (animal and plant tissues) and industrial materials (polymers, membranes, etc.).
- High-quality UT sectioning by UM helps in acquiring high-quality TEM micrographs.
- The ultramicrotomy procedure involves laborious sample preparation prior to the final sectioning of samples.

Reference

Latta H, Hartmann JF (1950) Use of a glass edge in thin sectioning for electron microscopy. Proc Soc Exp Biol Med 74(2):436–439

5 Ultramicrotomy: Instrumentation and Techniques

This chapter explains the practical details of successfully sectioning embedded specimens for TEM. It describes one of the most promising ultramicrotomy techniques that aims to retain and represent traditional ultramicrotome (UM) utility. Special light is given to the parts of the process that are the key to the successful sectioning of any biological specimen. Ultramicrotomy has been an essential skill for electron microscopists since the first light of TEM, although all the steps of ultramicrotomy have been exceptionally challenging. However, skilled microtomists have developed techniques to minister to this process over time. This chapter aims to assist beginners in ultramicrotomy to understand more about the principle and terms used along with its procedure and utility in successfully producing ultrathin (UT) sections and also the post-sectioning method, i.e., staining of UT sections to prepare the sample for TEM viewing. Achieving the desired UT sections requires special preparatory techniques to get quality transmission electron micrographs. The standard technique for ultramicrotomy involves many critical steps.

A. Haloi, *Essentials of Ultramicrotomy*,
https://doi.org/10.1007/978-981-97-4791-7_5

Learning Objectives from This Chapter

- Rapid evaluation of specimen block preparation and hand trimming of specimen block
- Preparation, use, and evaluation of glass knives
- Mounting of the water trough to glass knife edge and its use in ultramicrotomy
- Steps of machine trimming and its importance for semi-thin (ST) sectioning
- Specific trimming tricks if the specimen is too big
- Different steps of ST sectioning for light microscopy
- Staining of ST sections
- Final trimming of block face for UT sectioning and its importance
- Different features of diamond knife for ultramicrotomy
- Steps of UT sectioning technique
- Common errors during cutting procedures
- Washing and cleaning of diamond knife
- Different types of grids and types of film to support the grid
- Staining of UT sections and stain preparation
- Storage of prepared grids
- Negative staining and its preparation
- TEM viewing and reference micrographs

Ultramicrotomy Procedure

The specimen block is prepared through the steps: (i) primary fixation, (ii) washing with buffer, (iii) post-fixation, (iv) washing with buffer, (v) dehydration, (vi) clearing, and (vii) infiltrating and embedding to achieve the polymerised block (discussed in Chap. 2). Polymerised block can be capsulated (if no specimen orientation is needed) or flat block. The blocks are removed from the beam capsule or flat embedding moulds are hand-trimmed to remove the excess resin part surrounding the specimen to produce a pyramid form at the tip of the block where the tissue/specimen is located. Single-edged blades or clean razor blades do this activity with a block trimmer. The sides of the pyramid should not be too steep as it reduces the stability of the block while cutting sections and during machine trimming of blocks (Figs. 5.1, 5.2, 5.3, 5.4, 5.5, 5.6, 5.7 and 5.8).

At this point of trimming, care should be taken not to touch the specimen area, and it is necessary to keep some area but very little resin around the specimen. Leaving some resin surrounding the sample is needed to preserve the whole specimen area for ST sectioning before processing for UT sectioning. Fine trimming or machine trimming would be done in UM with the help of a Glass knife (Figs. 5.9 and 5.10).

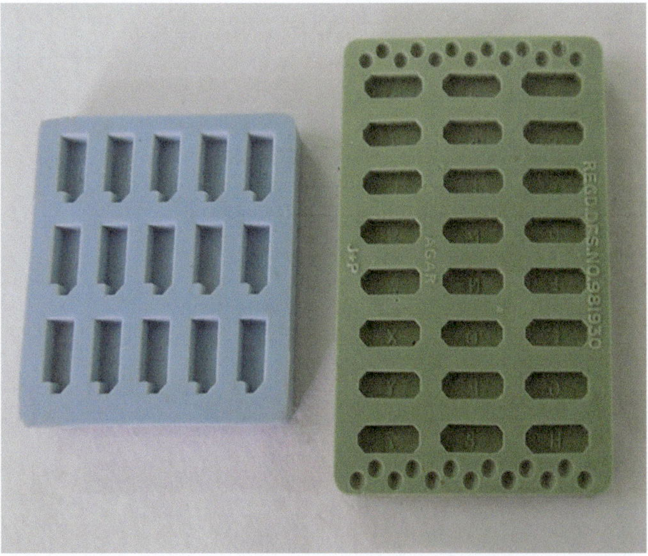

Fig. 5.1 Flat embedding moulds

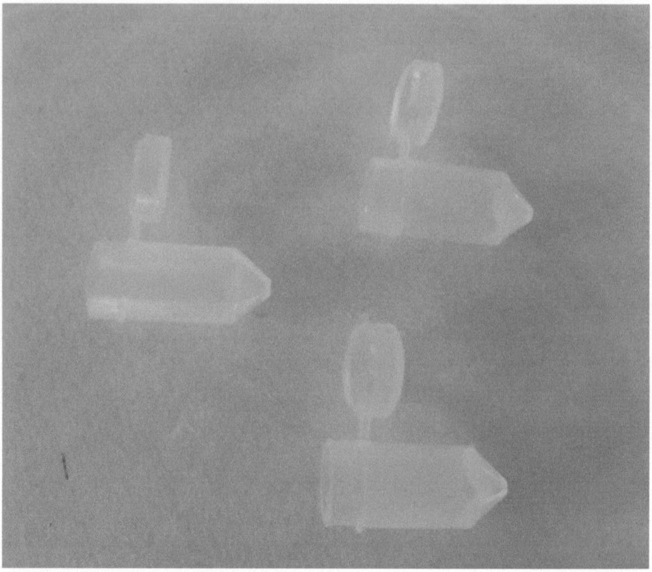

Fig. 5.2 Beam capsule embedding moulds

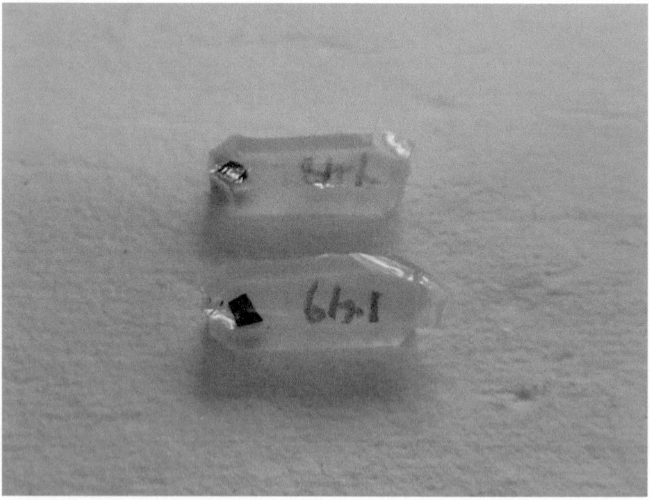

Fig. 5.3 Polymerised flat blocks

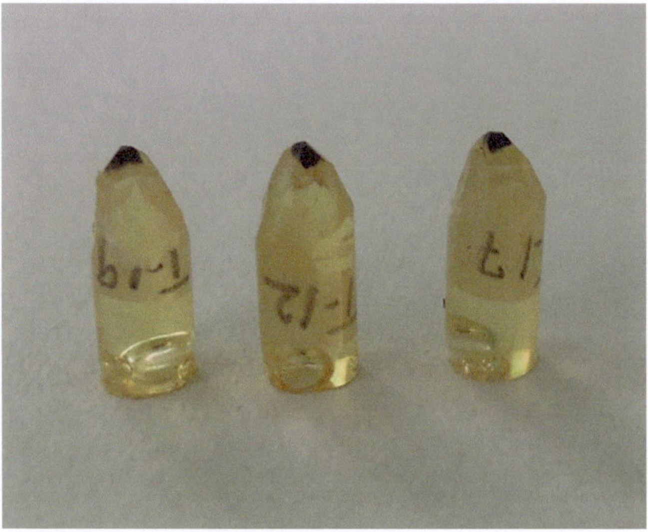

Fig. 5.4 Polymerised capsulated blocks

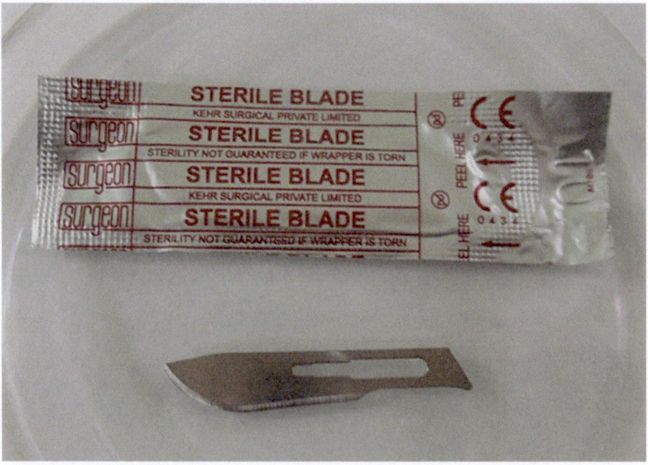

Fig. 5.5 Razor blade

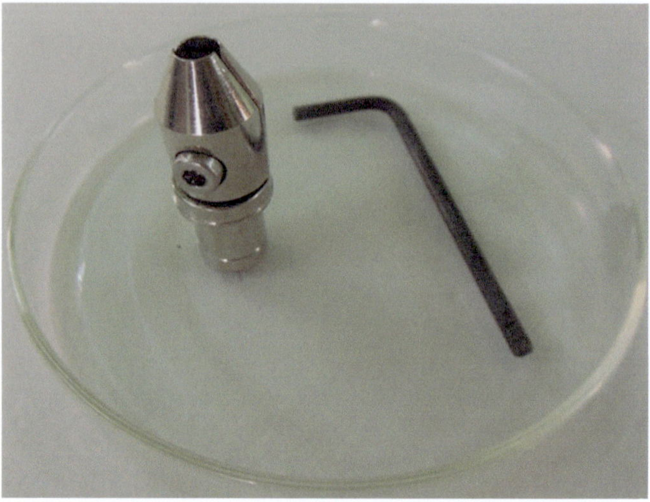

Fig. 5.6 Chuck Holder and Allen's key

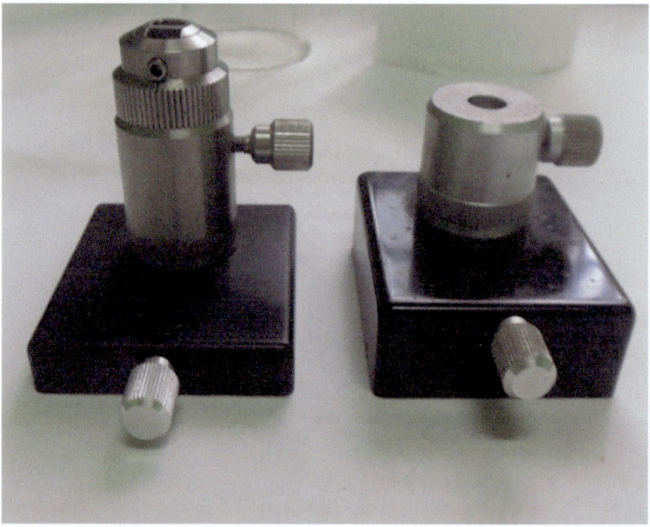

Fig. 5.7 Block trimmers

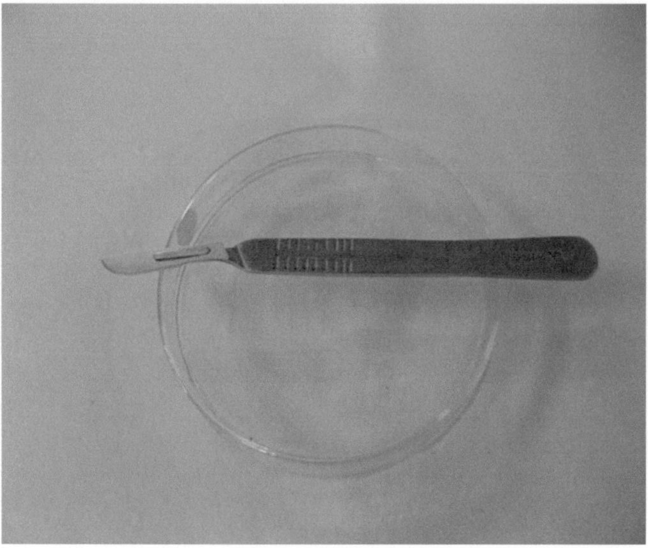

Fig. 5.8 Knife

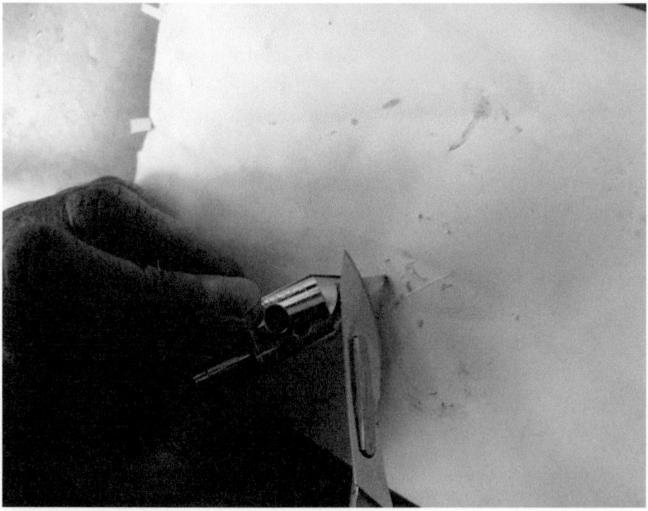

Fig. 5.9 Hand trimming

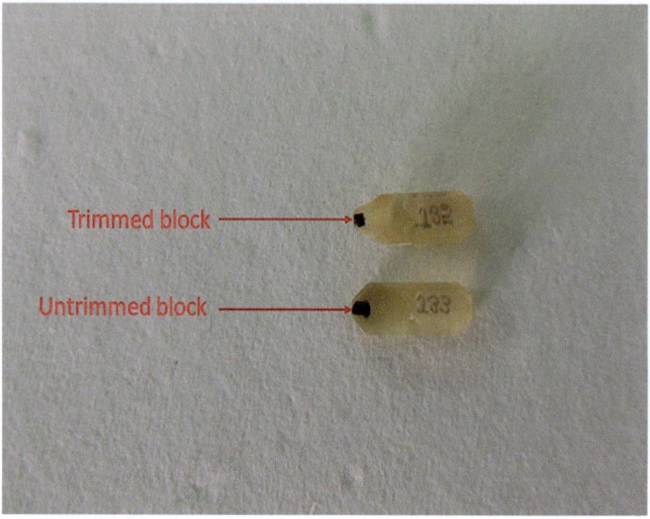

Fig. 5.10 Trimmed and untrimmed blocks

Glass Knives

Glass knives are used for machine trimming of the polymerised block, ST sectioning, and UT sectioning. Glass knives are prepared by special process glass strips and with a glass knife maker. Generally, glass strips are available in three types of thickness, i.e., 6 mm, 8 mm, and 12 mm, and in 200 mm and 400 mm lengths. To make glass knives, the glass strip has to break into the square first and then the glass knife from the square. The steps to produce good edged glass knife are as follows:

Steps to Produce Glass Square from Glass Strip

- The glass strips are washed with good quality mild detergent, rinsed with running water, and then dried in an incubator or left for air drying. To make a square, the strip should be adequately cleaned, as dirt or dust on the knife will destroy the cutting edge of the knife. To avoid any cut and other types of injury, the operator has to wear rubber gloves while handling the glass

Fig. 5.11 Washing a glass strip

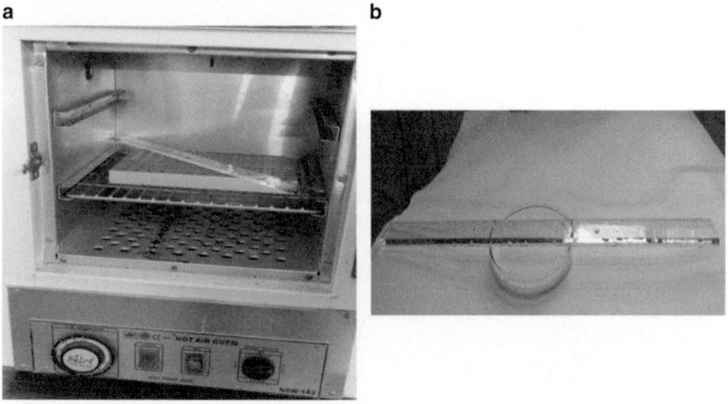

Fig. 5.12 (**a**): Drying of glass strip in an incubator. (**b**): Air drying of the glass strip

strip and a soft sponge can be used to clean off any residue on the glass strip. Final rinsing can be done with distilled water, and a good quality lint-free laboratory tissue can be used to dry the strip. Alternatively, the glass strip can also be left for air drying (Figs. 5.11 and 5.12).

- Before breaking the glass strip into a square, careful examination of the glass knife maker is needed to make sure that the

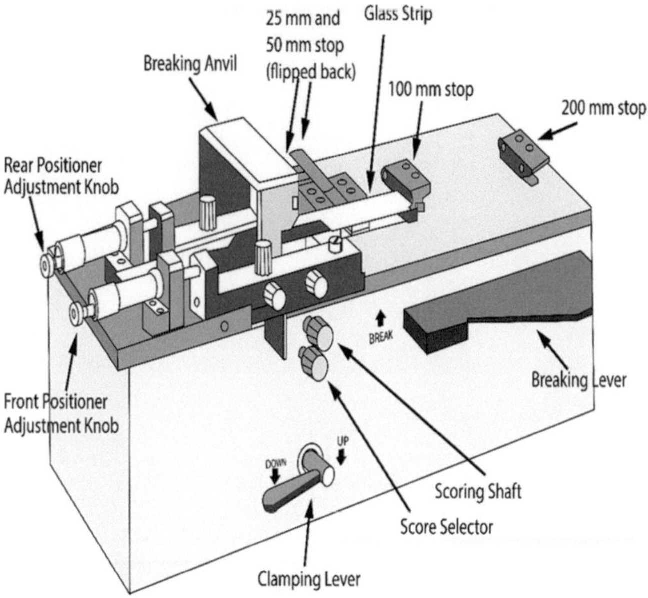

Fig. 5.13 Description sketch of glass knife maker

breaking lever is in the down position, the clamping lever is in the up position, the score selector is in number 1, the scoring shaft is all the way in (scoring wheel behind the glass strip) (Figs. 5.13, 5.14 and 5.15).

- One side of the glass strip is fine, and the other side is serrated length-wise. So now, while placing the glass strip on the instrument, the fine edge has to be down, and the serrated side has to be up in the glass knife maker. Now, the freshly cleaned glass strip should be placed on the top of the glass knife maker in such a way that its right edge stops against the 200 or 100 mm stop, depending upon the length of the strip. At this movement, the glass strip is firmly against the rear position, so the score will be strictly at a 90° angle. The glass strip should be held at this position with the help of the operator's hand (Figs. 5.16 and 5.17).

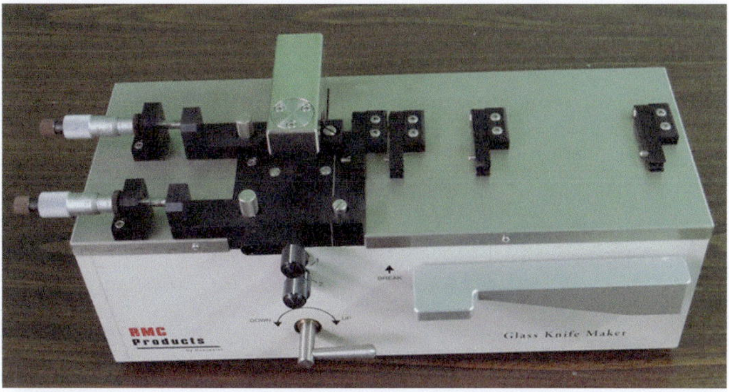

Fig. 5.14 Top view of glass knife maker

Fig. 5.15 Examine that everything is in position

- The glass strip has to be tightened in this position by rotating the clamping lever counterclockwise until firm contact is made. The scoring shaft has to be pulled forward towards the operator with a smooth and even stroke to make a score and leave in this position. This is the position where the glass strip should break (Figs. 5.18 and 5.19).

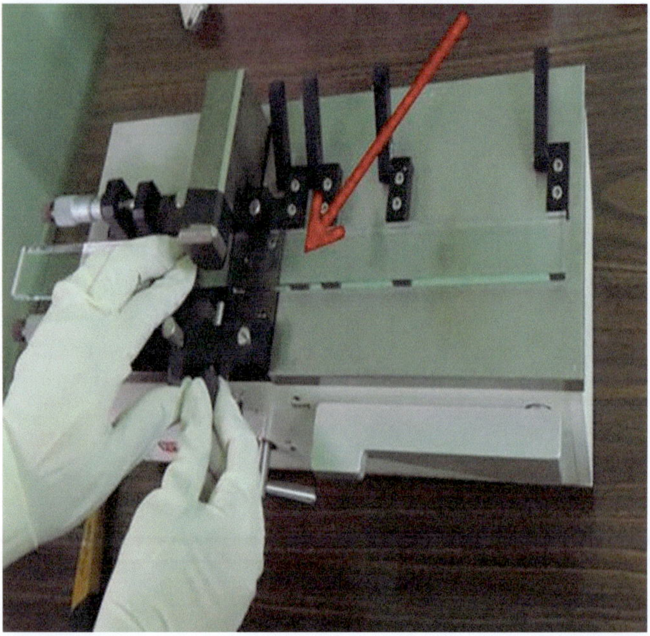

Fig. 5.16 Placing a glass strip on the glass knife maker

- To break at this point, the breaking lever should raise upward to apply pressure at that previously scored position. Faster breaks will generally yield good edges of the knife. This requires some practice. At this point, the glass strip will break in half.
- After breaking the glass strip, lower the breaking lever and lift the breaking arm from the glass strip by rotating the clamping lever clockwise. Carefully pull the freshly broken halved glass strip away to avoid any damage to the broken edge. Now, the scoring shaft has to push back inwards to prepare for the next break (Figs. 5.20 and 5.21).
- This method will break the glass strips into half, and each successive strip will be cut in half until the glass squares are produced. In this method, each break is as nearly perpendicular as

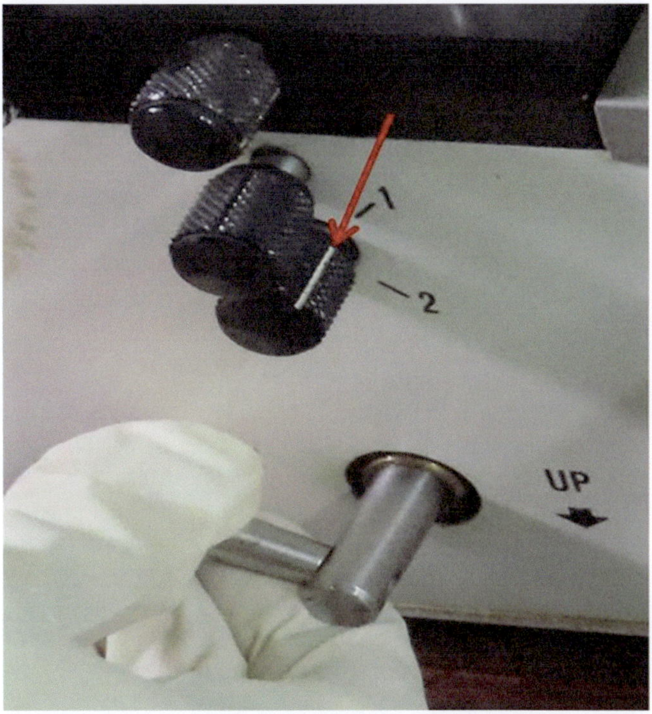

Fig. 5.17 The score selector is in number 1

possible and glass strips are breaking symmetrically because the glass strips break precisely in half using the appropriate stop, i.e., 100 mm, 50 mm, and finally, 25 mm so that all the breaks are balanced. There is another method where the squares are prepared starting from one end. In this case, unequal pressure is applied on each side of the score due to the greater weight of the longer side of the strip. So, the edges are not always perpendicular in this case. The glass squares are 25 mm (Figs. 5.22, 5.23, 5.24, and 5.25).

- As we go on breaking half of the strip, the 100 mm, 50 mm, and 25 mm stops should flip down to place the strip in a tight position. The right edge must be against the stop whenever we

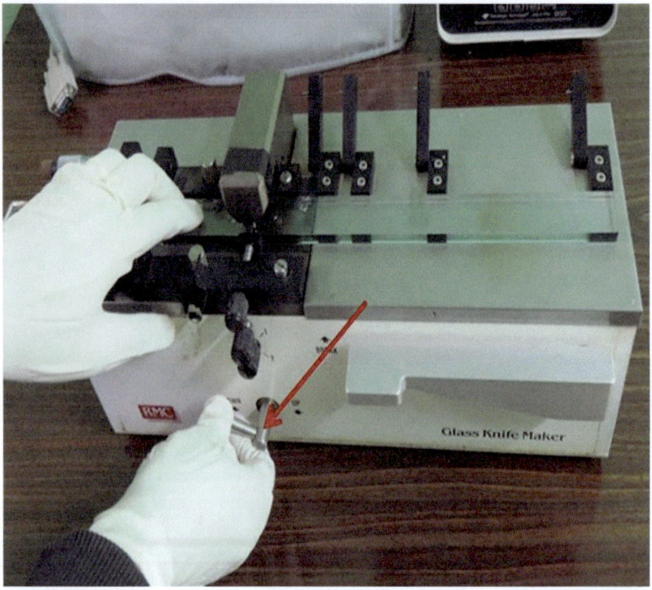

Fig. 5.18 Tightening the glass strip

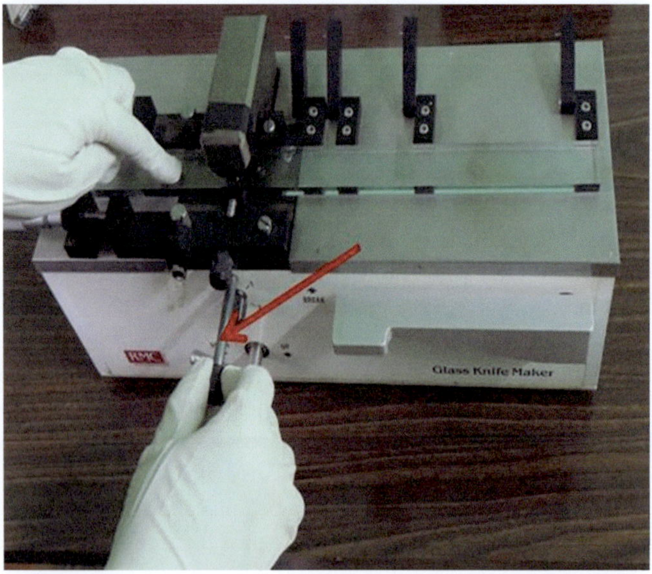

Fig. 5.19 Pulling the scoring shaft

Fig. 5.20 The breaking lever is raised upward to break the strip

Fig. 5.21 Lowering of the breaking lever

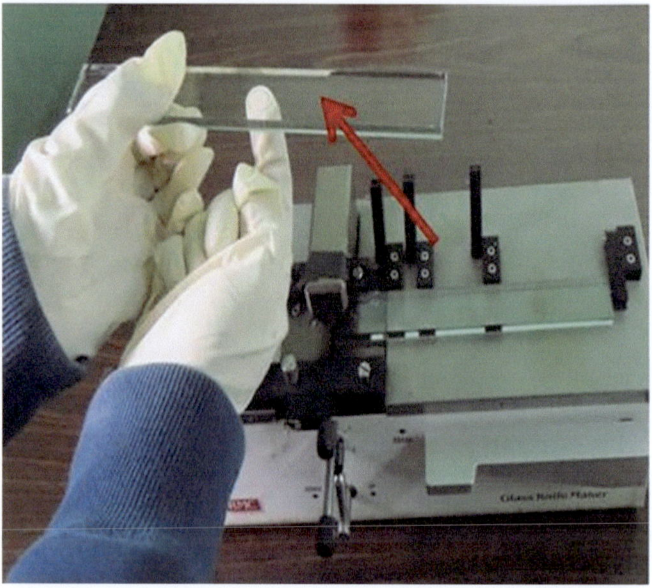

Fig. 5.22 Glass strip is brokeninto half

place it on the instrument. The last square through this process may not be useful as the left end of the strip will probably only be an exact square if the original glass strip is 400 mm long. The surface of the glass knife maker, mainly the top part, should be brushed off every time after the glass breaking operation with the help of a pointed brush to remove and clean all glass chips, and we should not brush the glass chips into the scoring shaft slot. This activity is to avoid uneven pressure and a poor-quality break, meaning the breaking edges may not be fine or may fracture edges, too (Figs. 5.26 and 5.27).

Making Knives from the Squares

Two usable glass knives can be produced from each glass square. This is done by scoring the glass squares nearly precisely

Fig. 5.23 Preparing to make a glass square from one end of the half-strip

diagonally. This alignment will produce a knife edge to one side of the top corner and the other side of the bottom corner. If the score is precisely diagonal, the glass square will break through the top and bottom corners, and these knife edges will be useless. So, the scoring needs to adjust to form good knife edges in either the upper left or the lower right corners and vice versa as the glass knife maker scores and breaks symmetrically.

Steps to Produce Glass Knives

- The scoring wheel has to be behind the glass square before placing the square onto the knife maker, which means the scoring shaft should be pushed all the way in, and the score selector should rotate to number 2 (Figs. 5.28 and 5.29).
- Although the knife edge can be made from both freshly broken sides and from the side broken by the manufacturer, it is

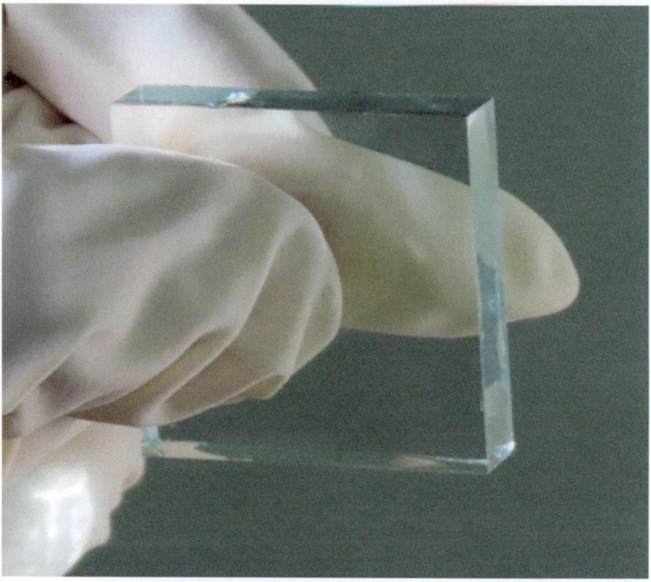

Fig. 5.24 Freshly prepared glass square

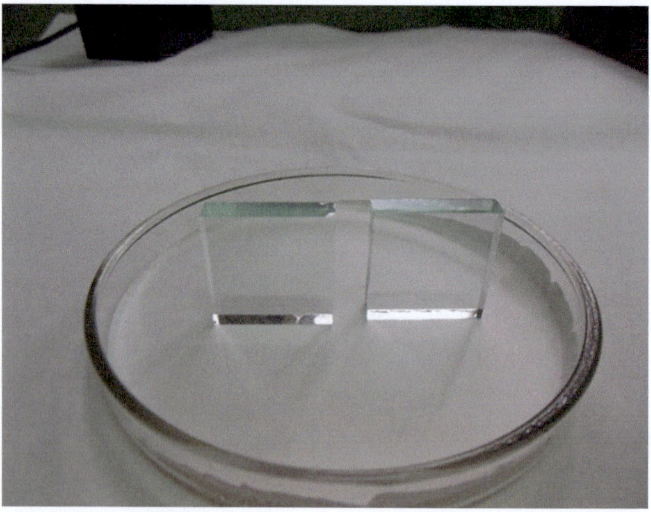

Fig. 5.25 Glass squares

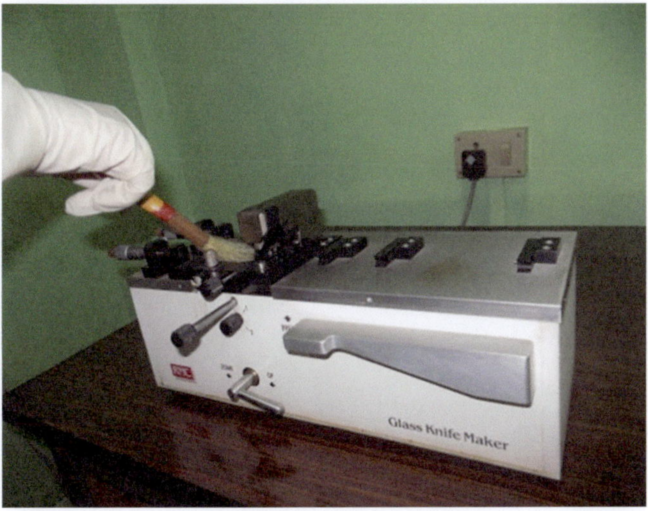

Fig. 5.26 Brushing off the top part of the glass knife maker

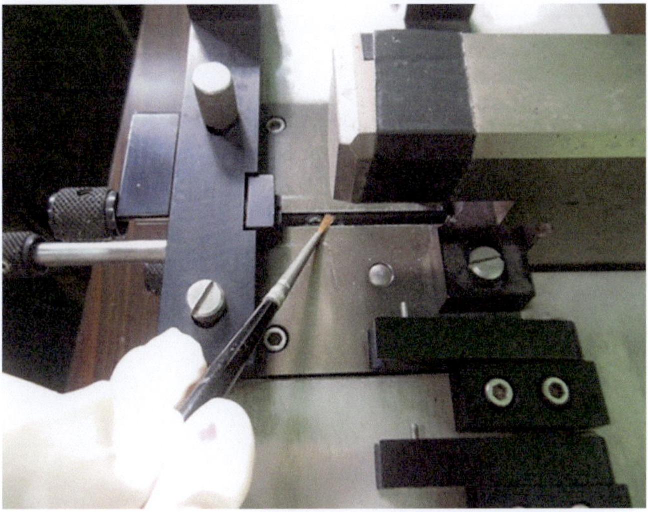

Fig. 5.27 Brushing off the scoring shaft slot with a pointed brush

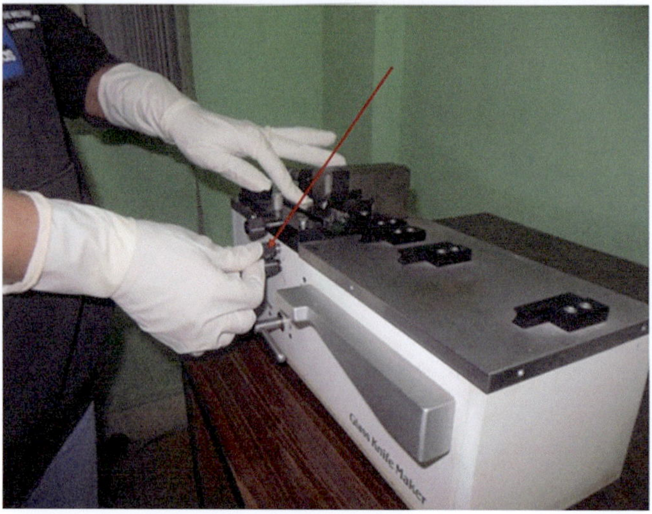

Fig. 5.28 Scoring wheel is pushed behind to the original place

Fig. 5.29 Score selector is in number 2

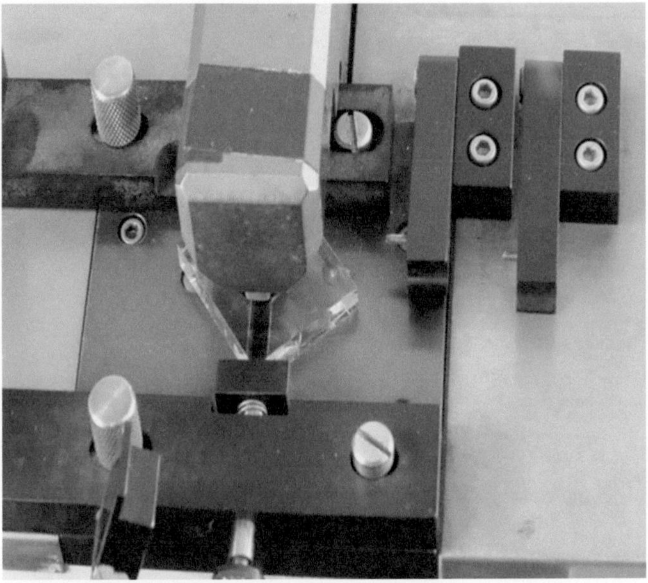

Fig. 5.30 Securing the glass square between the notches of the front and rear position

preferable to make the knife edge from the sides produced by the manufacturer because these edges are very nearly perpendicular and have minimal stress lines. Now, the glass square has to be positioned and oriented on the knife maker in such a way that the edges of the knife will be formed. Front positioned retract knob has to pull back. The glass square must be inserted by placing the far corner of the square in the notched space of the rear position, and the operator side corner should be positioned by the notch on the front position. Now release the spring-loaded retract knob to secure the glass square between the notches of the front and rear positioned (Figs. 5.30 and 5.31).

- To break the glass square on the scoring, lower the breaking anvil by rotating the clamping lever counterclockwise and to

Fig. 5.31 Diagrammatic representation of Glass square is positioned

get an even score, pull the scoring shaft towards the operator with a smooth motion to make the score and then raise the breaking lever to exert pressure on the square and break it to two triangles (Figs. 5.32, 5.33, 5.34, 5.35, 5.36 and 5.37).

- Turn the clamping lever clockwise to raise the breaking anvil after the glass square breaks. The front positioned retract knob should pull back with one hand and carefully remove the two halves of the glass square or the freshly prepared triangle from under the breaking arm with the other hand. Through this process, two usable knives are achieved. One sharp edge, which is the cutting edge, is formed in one triangle, and the other side of the sharp triangle is blunt or has a shaft at each knife. Since only one knife edge can be formed at each corner, the shelves must be about 0.5–1 mm wide (Figs. 5.38 and 5.39).

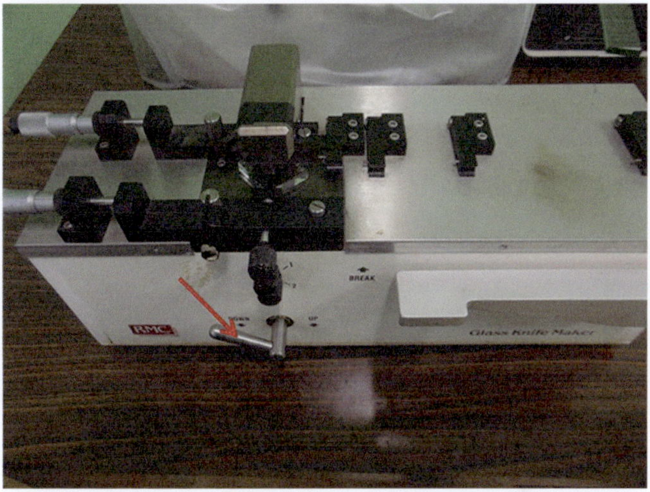

Fig. 5.32 Lower the breaking anvil

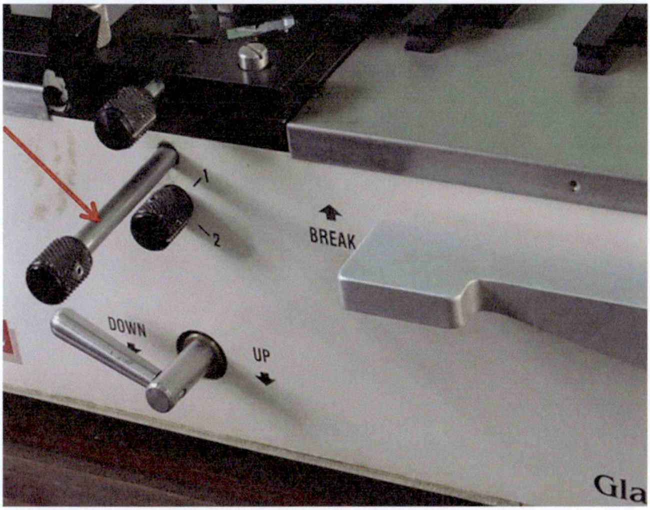

Fig. 5.33 Pulling the scoring shaft

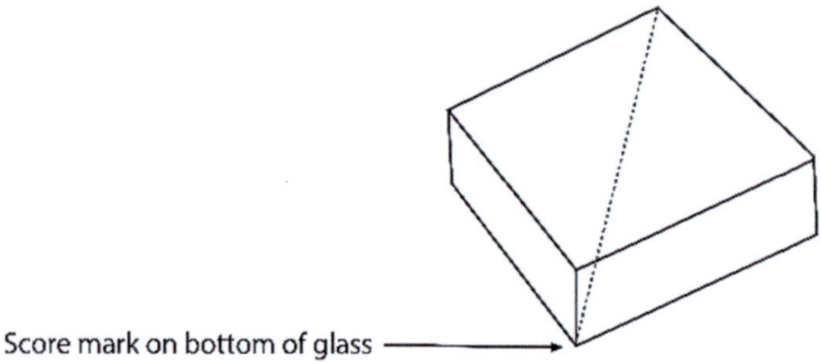

Fig. 5.34 Diagrammatic representation of score mark on the bottom of the glass square

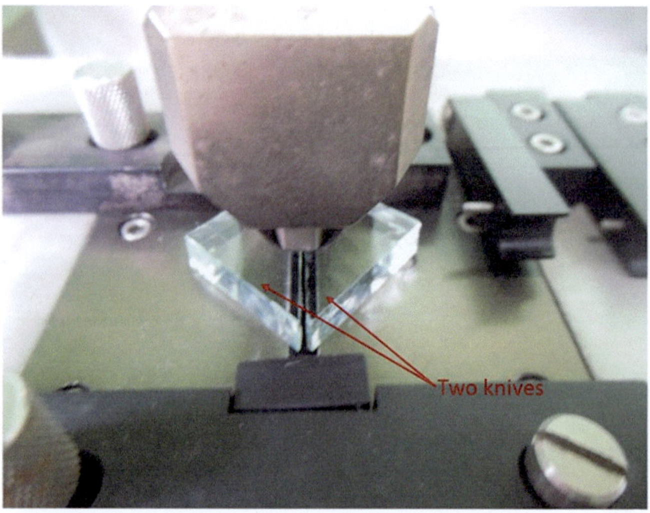

Fig. 5.35 Two triangles (glass knives) achieved by breaking the square on the scored mark

The quality of the knife edge largely determines the quality of UT sections. Glass knives produced by glass knife makers have reproducible shape and quality; however, the usable length of the knife edge for sectioning may vary slightly in different knives, and this has to be checked by evaluating the knife edge before use.

Fig. 5.36 Two usable glass knives from one square

It can be done by the knife evaluating system, which comes with the instrument (ultramicrotome).

Steps to Evaluate

Place and lock the freshly prepared knife in the knife evaluation system. The back of the knife should push firmly against the vertical back plate. Only 45° knives will rest on the horizontal base plate of the knife evaluation system. Place the knife evaluation system on ultramicrotome and switch on the flex optics system. The knife evaluation system and the glass knife should be in front of the knife stage so that the beam from flex optics transmits into the mirror under the knife. Now, pull the microscope stand towards the microtomist and focus on the knife edge by

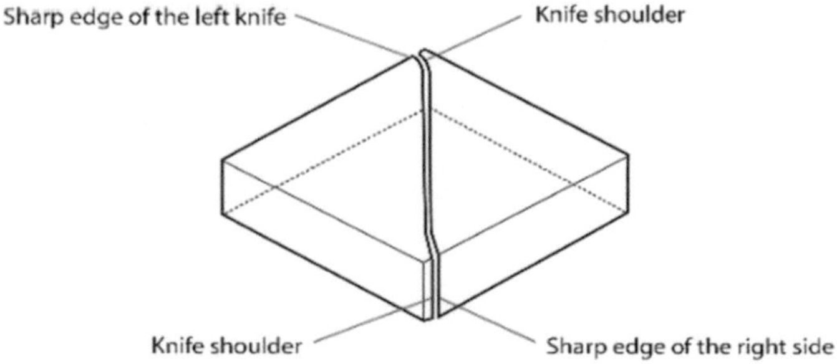

Fig. 5.37 Diagrammatic representation of sharp edge and knife shoulder

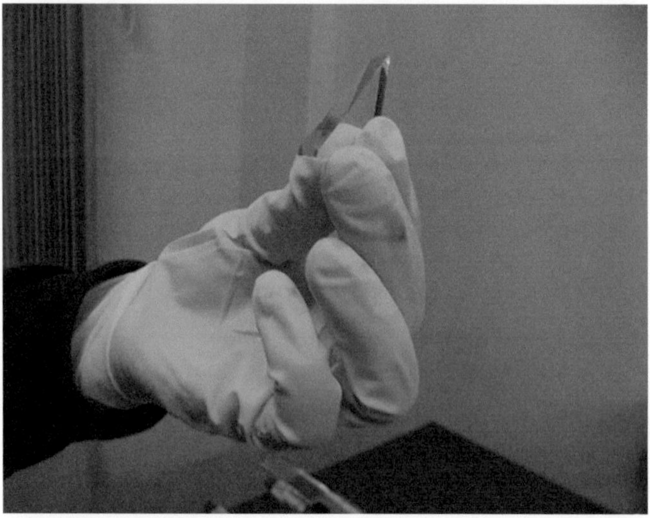

Fig. 5.38 One half is one glass knife

adjusting through the focus knob, zoom knob, microscope tilt FB (forward–backward) knob and stereomicroscope travel knob. Increase the magnification to have the entire view on the edge. The screw on the right side of the knife evaluation system should turn to rotate the mirror under the knife until a bright reflection is seen at the tip of the knife edge. The magnification may further

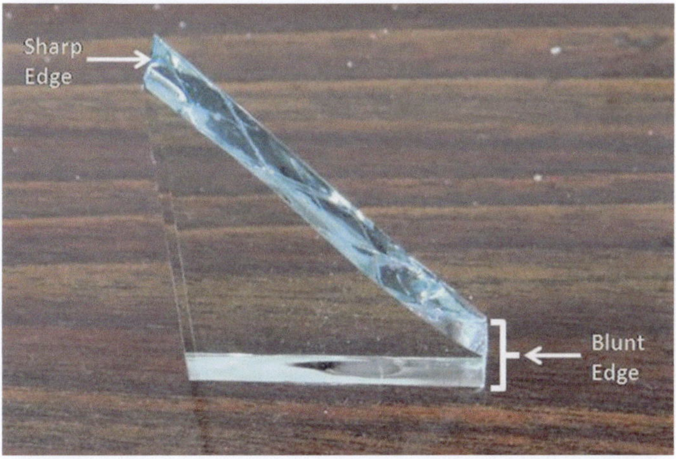

Fig. 5.39 Showing the blunt and sharp edges of a glass knife. *See also* Fig. 5.40

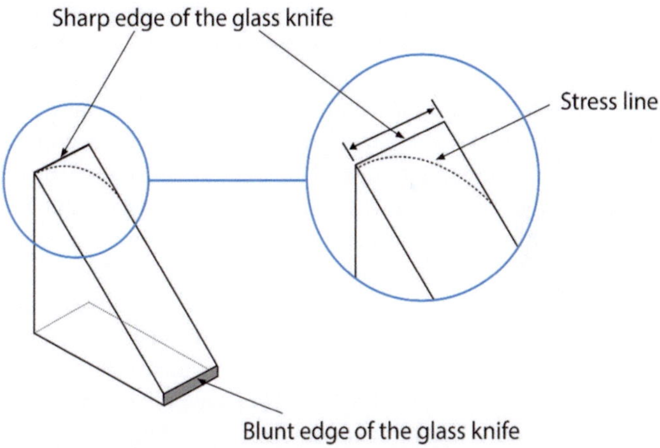

Fig. 5.40 Diagrammatic representation of sharp edge, blunt edge and stress line of glass knife

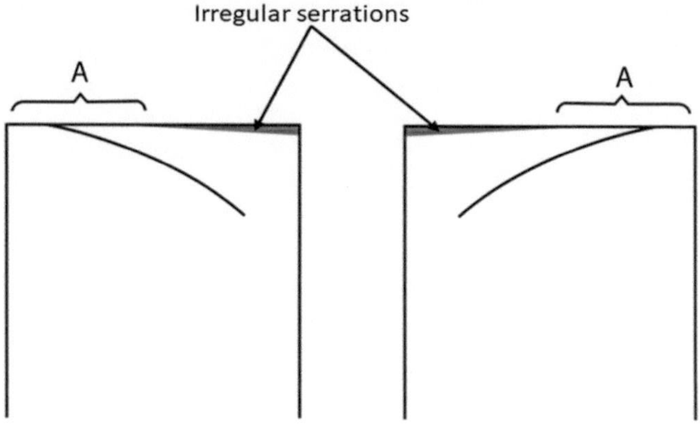

Fig. 5.41 Diagrammatic representation of the good area "A" and irregular serrations of the knife edge

increase for a more critical evaluation. If the stress line emerges from the left corner, as shown in Fig. 5.41, this side of the knife edge, says "A", will be smooth and without any defects, and this good area can be used for ultrathin sectioning. On the other hand, the other side of the knife edge, i.e., the right side, may have irregular serrations, which can be used for semithin sectioning and also for trimming blocks. These irregular serrations will increase in intensity towards the right. As the picture depicts, if the stress line of the knife edge emerges from the right corner, the good area "A" is on the right side of the knife. Glass knives should be prepared just before use as the freshly broken knife edge is suitable for sectioning due to the molecular arrangement in the glass and the knife edge can also get contaminated easily if it rests for a longer time. Perhaps the squares or rhombi can be prepared and stored in advance. Once the glass knife is prepared in the knife maker, it should be removed from the knifemaker with extreme care to prevent the knife edge from any damage, and it is advisable to examine or evaluate the knife edge prior to sectioning (Figs. 5.40 and 5.41).

Mounting a Water Trough to Glass Knife Edge

Glass knives are used for machine trimming of blocks and also for cutting ST and UT sections. For machine trimming of blocks, glass knives can be used as they are, and for cutting semithin and ultrathin sections, the trough has to be fixed to the glass knife edge to produce a boat where the sections can be collected later. Apart from the readymade trough (Truf), plastic tape is also available for this purpose. Usually, disposable plastic Trufs are used, which is the quickest and cleanest type of trough. Trough is designed to fit the glass knife sizes, and each turf is packed separately, completely clean, and ready to use. These Trufs give good water reflection, so the sections are clearly visible on the surface of the trough liquid during sectioning. Trufs are fixed to the glass knife edge with dental wax and dispensed from a multiplate. Commercially available nail polish can also be used directly to fix the Truf to the edge of the glass knife in place of dental wax (Figs. 5.42, 5.43, and 5.44).

Steps

- The multiplate has to be switched on 20 min before use. The spatula of multiplate should be placed in the well to warm it, and the dental wax has to be added to the hot deck and well. The freshly prepared glass knives must be placed on the hot deck to warm it for about 30 s. The cutting edges of the glass knives should protrude over the edge. The plastic ready-to-use Trufs should be placed on the wax dispensing deck for about 10 s. The left edge of the Trufs is always broader than the right edge so that it can fit into the groove of the wax dispensing deck (Fig 5.45).
- The waxed Truf has to be directly mounted onto the warm knife by aligning the upper edges with the knife edge and holding in that position for a few moments with the hand, which allows the Truf to stick to the glass knife. The knife's edge should not be touched with any object throughout the process (Fig. 5.46).

Fig. 5.42 Plastic troughs (Truf) for fitting to glass knives

- After the glass knife with the Truf cools down, seal around the edges (sides) of the Truf with the melted dental wax from the well so that there will be no leakage of the Truf liquid while sectioning. Allow it to cool down for a few moments before use (Figs. 5.47 and 5.48).
- The glass knife with Truf can be used for one-time sectioning only to prevent contamination. After the sectioning of one block is complete, the used knife with Truf should be safely disposed of. However, if the knife edge is still fine, this glass knife can only be used to trim the block by removing the Truf from it.

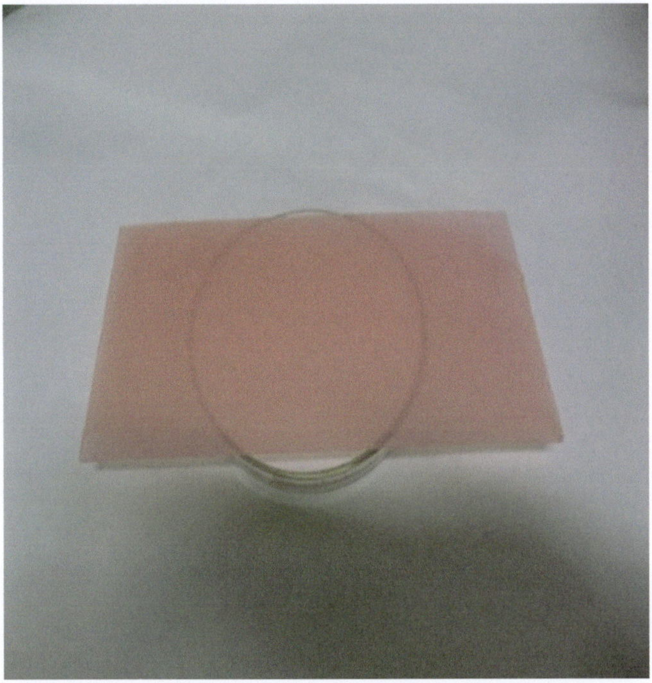

Fig. 5.43 Dental wax

Machine Trimming for Semithin (ST) Sectioning

As in hand trimming, the face of the block, which contains the specimen area, is not smooth; machine trimming is followed up to smoothen the surface of the block face. The purpose of machine trimming is to make a trapezoid, square, or rectangle shape of the specimen on the face of the block. The specimen's smooth face and parallel edges will give serial sections while sectioning, and one section will adhere to another to form a ribbon. While trimming to a trapezoid, square or rectangle shape, the surrounding resin of the sample is needed to remove as much as possible and may keep the whole specimen for ST sectioning with a little surrounding resin. So that all required information about the tissue/

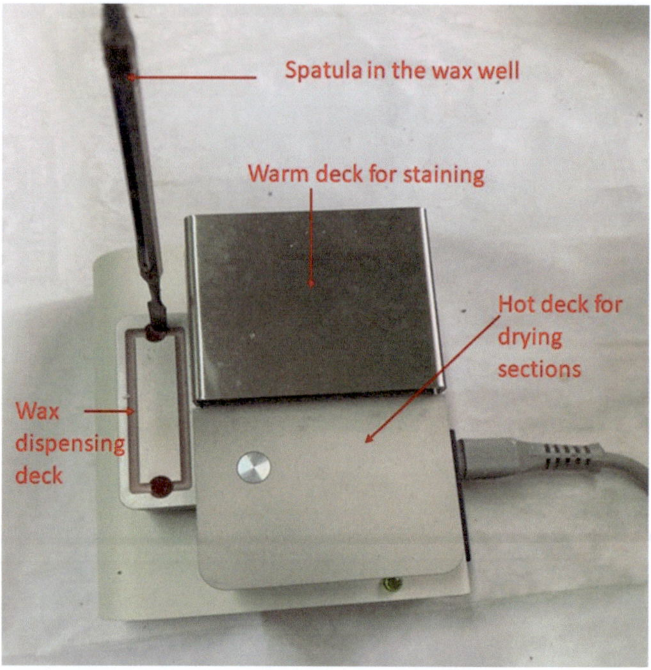

Fig. 5.44 Multiplate

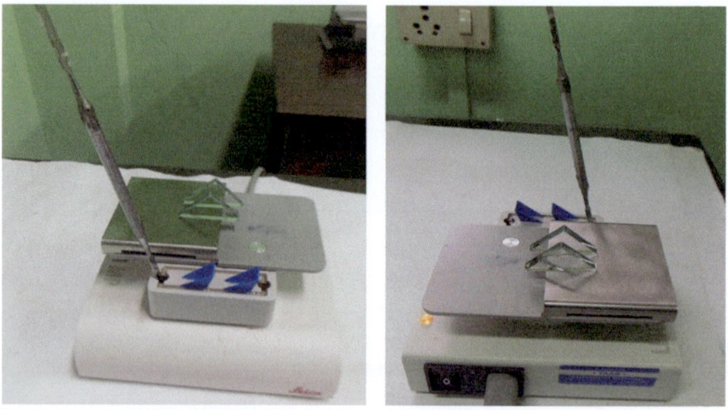

Fig. 5.45 Glass knives and Trufs are placed on the multiplate

Fig. 5.46 Mounting a waxed Truf on the warm knife

specimen is preserved to examine under the light microscope. At this stage of trimming, the cutting face of the block is typically trimmed to trapezoid form if going for ST sectioning before processing for UT sectioning because, in that case, while examining the ST sections under a light microscope (LM), it is easy to locate the particular area of interest on the section for final trimming of the block face. The status of the block, such as the consistency of the embedding medium, the presence or absence of holes in the specimen block and whether the area of interest is present, can also be examined under an LM at this stage. The trapezoid face of the block also facilitates serial sectioning. A trapezoid, about 30 μm in length, 50 μm in base width and 15 μm in height, with a parallel top and bottom, can produce one long continuous ribbon

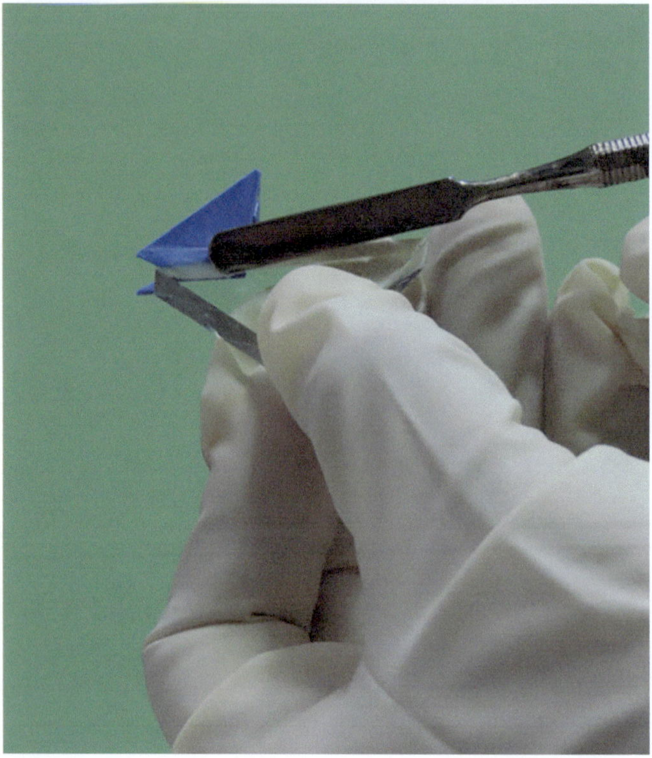

Fig. 5.47 Sealing around the edges

of many tightly linked sections. Achieving serial sections though a tedious job makes it possible to pick up and analyse the specimen at different depths. If the ribbon is too long, about 200 sections on one ribbon, then it must be subdivided into shorter segments to ease the picked-up process onto grids.

Steps for Machine Trimming (Manually)

- For fine trimming, a glass knife has to insert at the place of the knife holder of the knife stage (refer to Fig. 4.3 from the previous chapter) of the UM and lock it firmly with the knife clamping knob. Specimen block should place on the block holder and

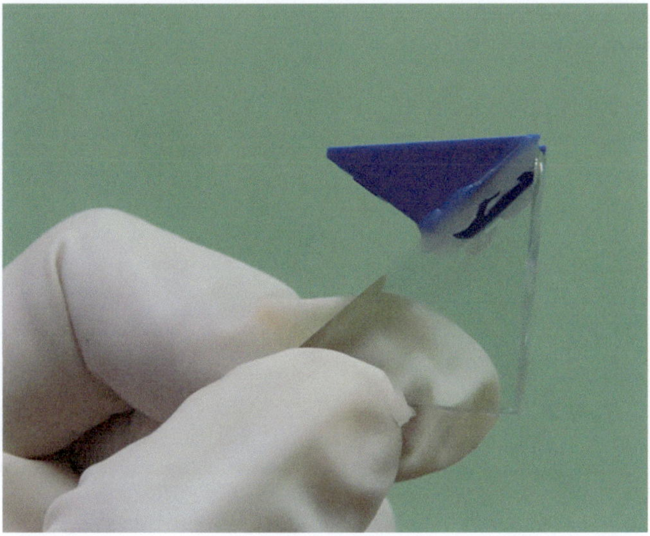

Fig. 5.48 Prepared glass knife with Truf

tighten with a specimen locking screw, and the block holder should fix at the arc segment mount of the UM and secure with arc adjustment knob. While fixing the block to the holder, 1/3 of the block should be in and the specimen side out to give good stability to the block throughout trimming and cutting. The height of the glass knife and block face may be adjusted in such a way that both are about to touch but still apart from each other by adjusting the position of the knife with the lateral knife angle adjustment knob of the upper stage of UM and height by hand wheel (Figs. 5.49, 5.50, 5.51, 5.52, 5.53, and 5.54).

- Then, switch on the overhead diffuse light from the lighting system of the control panel. This light will allow the microtomist to see the block face and the knife edge through the stereomicroscope of the UM. Focus on the knife edge and adjust the block face and knife by using the focus knob, zoom knob,

Fig. 5.49 Glass knife placed on the knife holder

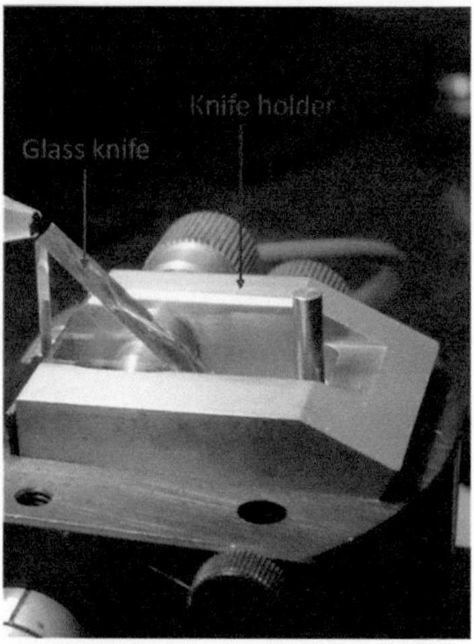

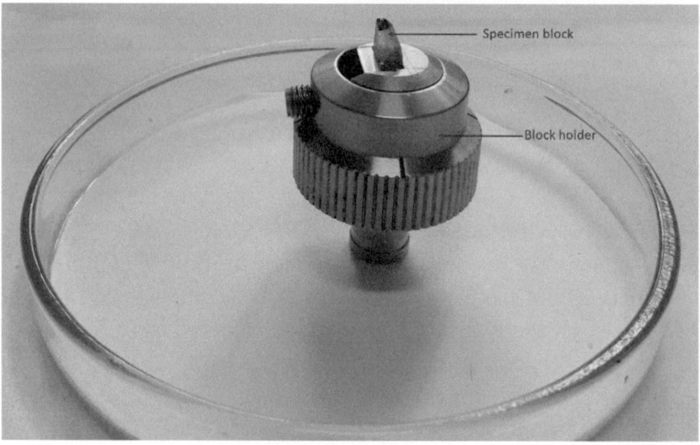

Fig. 5.50 Specimen block placed on block holder

microscope tilt FB (forward–backward) knob, and stereomicroscope travel knob (Figs. 5.55 and 5.56).

- Now, forward the glass knife towards the block face by using the coarse adjustment knob from the lower stage until the knife slightly touches the block face at the position/alignment of 0 from the upper stage and the arc segment mount and block

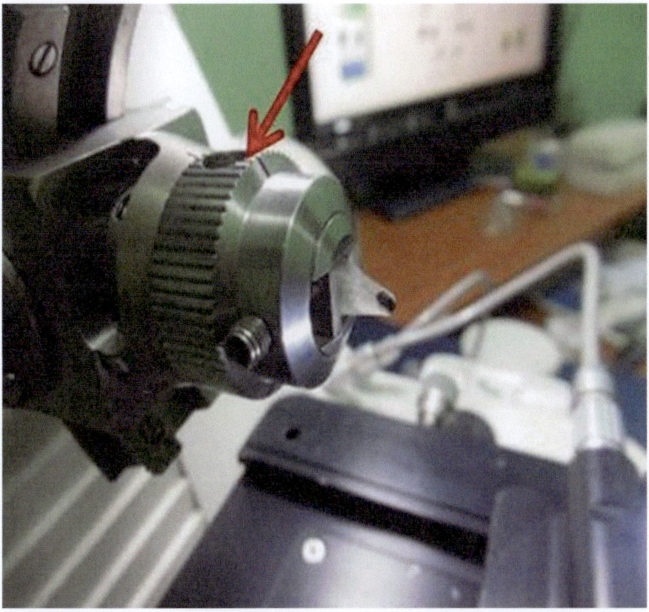

Fig. 5.51 Block holder fixed at arc segment mount

holder. Now, with the right hand, rotate the hand wheel up and down the specimen and forward the knife with the left hand (one stroke up–down and the other stroke forward the knife) until you reach the specimen and then trim to make a smooth face (Figs. 5.57 and 5.58).

- To trim the right side of the block, unlock the knife stage by losing the knife clamping knob and rotate the knife anticlockwise and fix the position at 0 and 35 from the upper stage and turn the block 1-degree right side to get a slanting edge of the block. If necessary, lateral position of the knife should be adjusted by using the required knob. Then, follow the previously mentioned step to reach to the specimen and continue trimming (Figs. 5.59 and 5.60).
- To trim the left side of the block, unlock the knife stage again and rotate it clockwise to bring it to the position of 0–35 from

Fig. 5.52 Adjusted the position of knife and specimen block face

the left side of the upper stage. Turn the block 1° left side to give slopping to the edge of the block face. Both the slopping side and the front face of the block are now ready (Figs. 5.61 and 5.62).

- To make two parallel sides of the trapezoid face, rotate the block holder 90° right side to bring it to a horizontal position and trim the side by putting the upper stage at 0–35. Similarly, to trim the left side, turn the knife stage to the left side at 0–35 and trim slices until the specimen is reached. Trimming of the block is now complete; at the end of it, we get the trapezoid face of the specimen. Return the orientation to the vertical or original position by turning the block holder. The block is now ready for ST sectioning (Figs. 5.63 and 5.64).

Fig. 5.53 Alignment of knife stage for trimming

The trapezoid face of the block is usually prepared from the whole specimen; however, in some cases, if the specimen is too big or the sample area is too large, it is not possible to take the whole area for sectioning and needs to remove some area of the sample; some specific trimming tricks have to follow where we can minimise the risk of cutting off required area from the specific sample as it is big in size and the scholar manually cannot provide the specific area because of difficulties in dissecting to small pieces.

Trimming of Parasite/Worm Sample and Intestine Sample

The periphery should always be included for a parasite specimen block or an intestine sample, as the microvilli are present in that area and are part of most of the study. The orientation of the sam-

Fig. 5.54 Showing diffuse light which enables microtomist to see the knife and specimen block area

ple is needed while preparing the block at the time of embedding the specimen in pure embedding media to keep the periphery area of the specimen, and flat embedding moulds should be used to polymerise this type of sample. While trimming this type of sample block, the periphery side should be kept, and the other side has to be removed to make a smaller face of the block for sectioning.

Similar trimming tricks like parasite or intestine samples have to be followed in the case of a leaf sample.

Trimming of Root Sample

Roots are usually round in shape, and if they are divided horizontally and vertically, one-fourth of them will generally represent the structure of the whole root. So, in trimming the root sample, one-fourth of it is taken to prepare the trapezoid face with a small

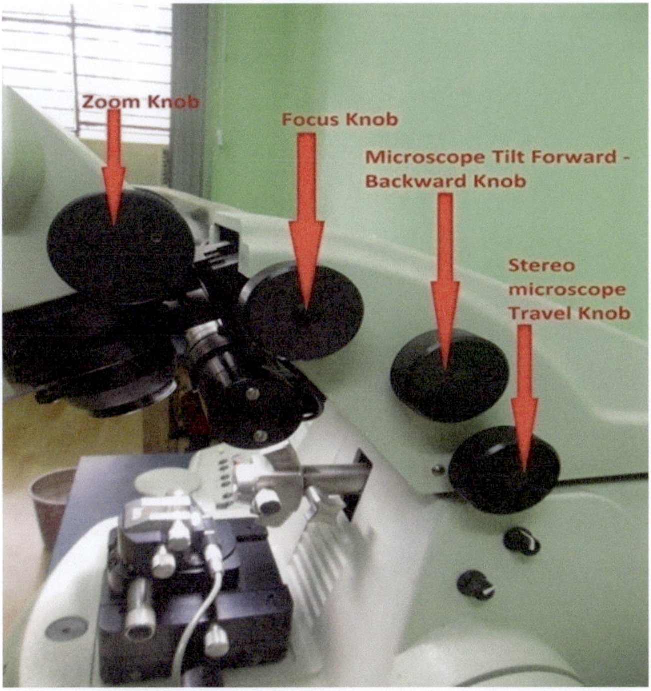

Fig. 5.55 Focus knob, zoom knob, microscope tilts FB (forward–backward) knob and stereomicroscope travel knob

quantity of resin intact from the outer side to include the periphery of the root, and the remaining part is to be trimmed off.

Trimming of Sperm, Ovary, Bacteria, and Virus Suspension Sample

In the case of these microscopic samples, the pallet of the suspension sample (after centrifuging) should be polymerised in the pure embedding medium in a microcentrifuge tube itself. After polymerisation, it may be removed from the embedding oven, and the pallet may be cut into multiple small pieces with the help of a laboratory saw. Then, the pallet pieces are re-embed separately in BEEM capsules or flat embedding mould to prepare blocks for

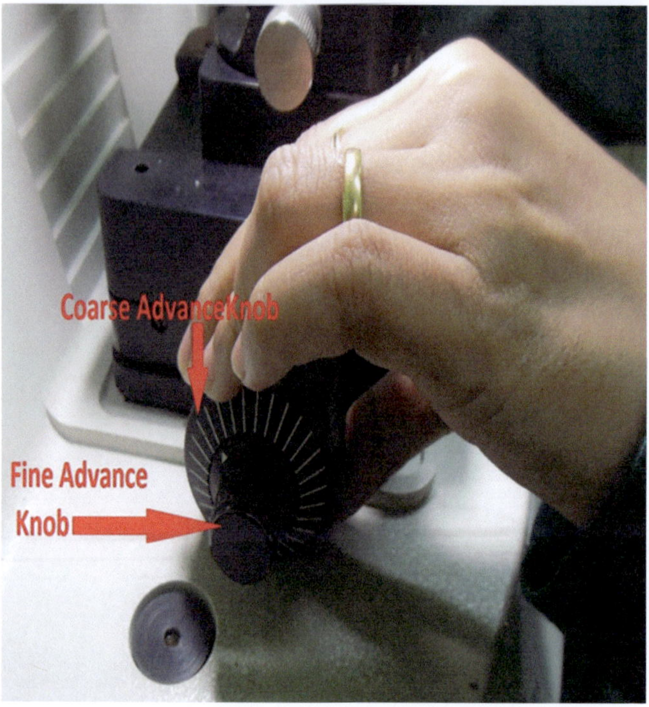

Fig. 5.56 Forwarding the glass knife towards the block face by using coarse adjustment knob from the lower stage with the left hand

sectioning. These blocks can be trimmed from any area of the specimen to get a trapezoid face.

ST Sectioning for Light Microscopy

Before processing for UT sectioning, ST sections of about 400–500 nm thickness are cut to examine the specimen/tissue under an optical microscope. ST sections are called survey sections in this case, enable us to determine the status of the block, like the state of the embedding medium (whether soft, medium, or hard), the fixation, dehydration, and infiltration of resins are

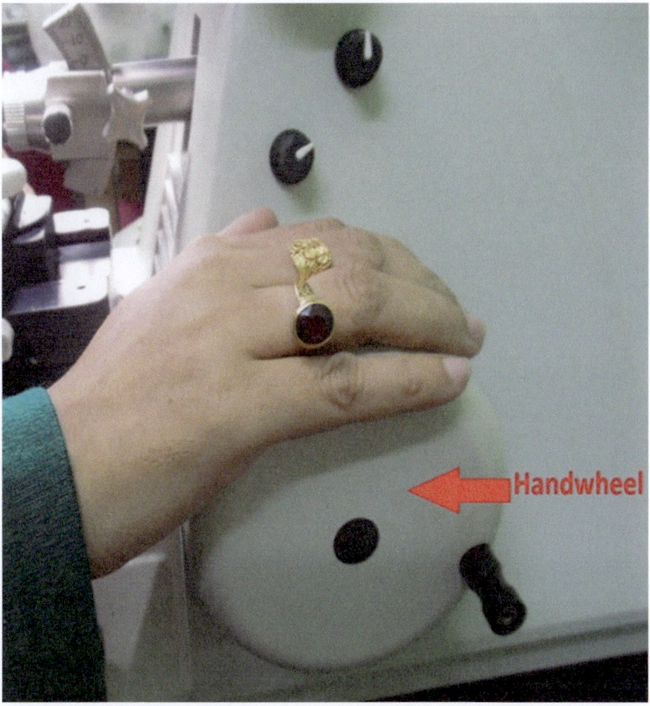

Fig. 5.57 Rotating the hand wheel to up–down the specimen with the right hand

proper or not (if any holes are present in the tissue), whether the area of interest is present or not. ST sections enable us to select the area for final trimming for UT sectioning.

Steps

The block should be at the position of original alignment while machine trimming, i.e., 0–0 from the arc adjustment knob and block holder. Glass knife which was used for machine trimming should remove and, in that place, freshly prepared glass knife with trough/boat to be placed, i.e., in the place of glass knife holder (refer Fig. 5.48). The selected area of the glass knife for cutting should align with the block face (trapezoid) in such a way that

Fig. 5.58 Magnified view of trimming of the front face of the block through a stereomicroscope

both of them are about to touch. This procedure is done manually. As the microtomist forwards the knife edge towards the block face, through the reflection of the backlight, the face of the block will shine, and as the knife approaches the block, the reflection will become gradually thinner until it appears a very thin line. The thin line means the knife edge and the block face are very near and about to touch, representing the distance between the knife edge and the block face. If needed, a viewing mirror can be used. The specimen arm's cutting speed usually varies from 0.5 to 2.0 mm/s, and to find the optimum cutting speed for different samples, one needs to experiment. However, the cutting speed should be set according to the consistency of the block (soft, medium, hard, or brittle) from the control panel (manual or computerised either). For a softer block, the cutting speed is set on the lower side; like-

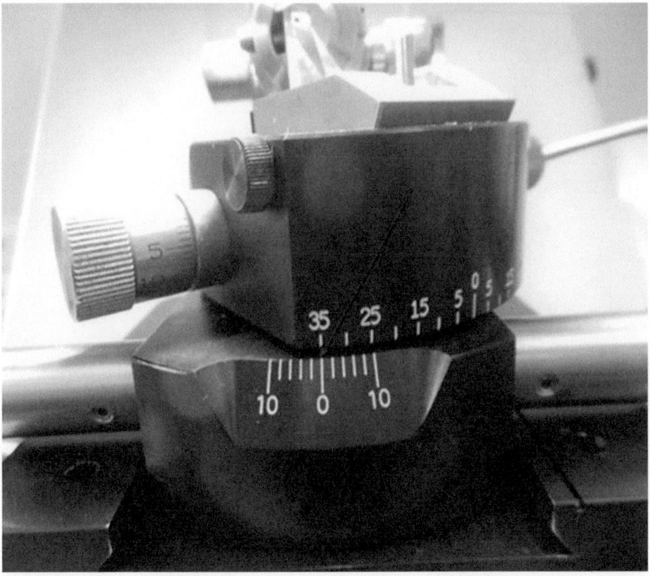

Fig. 5.59 Fixing the position at 0 and 35 from the right side of the upper stage

wise, a higher cutting speed is required for a harder block. The cutting speed is adjusted to about 1 mm/s in standard cases.

Then, turn on both overload diffuse light and backlight from behind, which illuminates the specimen from behind by depressing the appropriate button from the lighting system of the control panel. This light from above and behind the knife will enable the microtomist to see the knife's edge and the block face clearly and in a magnified view. The lighting system of UM also helps to set cutting windows, fills the boat, observes the sections, and picks up the sections throughout. After that, the cutting zone is set by turning the handwheel until the specimen is about 1 mm above the knife edge and then depressing the set upper W/handwheel button to set the upper limit of the cutting window at this position. Then, turn the handwheel until the specimen block face just passes the knife edge and depress the set lower W/handwheel button to set the lower limit of the cutting window at this position. The cutting

Fig. 5.60 Turning the block 1-degree right side to trim the side

window is set now. Within this distance of the cutting window, the sections will be cut at the given speed and thickness set; otherwise, the cutting arm's remaining part of the complete stroke will move at its own default speed. Then, section thickness needs to be set according to the requirement. Usually, between 400 and 500 nm, the semithin sections are cut. The section thickness button from the control panel displays in a nanometre scale, and there is a facility to press the "UP" arrow to increase the thickness and the "DOWN" arrow to decrease the thickness. One nanometre (1 nm) change can be made by depressing either of these "UP" or "DOWN" buttons once. If any of these two buttons are held for more than 10 s or continuously, the value will change steadily at a rapid rate (Figs. 5.65, 5.66, 5.67, 5.68, 5.69, 5.70, 5.71, and 5.72).

Filling up the knife boat is a critical step because more water on the boat will touch the face of the block, and the wet block face will not cut any sections. On the other hand, less water on the boat

Fig. 5.61 Fixing the position at 0–35 from the left side of the upper stage

will damage the knife edge as the cut sections will stick to the knife edge instead of floating on water. So, it is required to fill the knife boat to enable the cutting process properly. A syringe or water trough filler is used to fill the distilled water in the boat in a convex manner, which means putting a few more drops of water that will touch the knife edge but at the same time will not overflow. Sometimes, if the convex water level does not touch the knife's edge, the water level can be brought to the knife's edge with the help of a soft single hair brush or eyelash brush without touching the edge. At this point, the edge of the knife area will appear dark. With a syringe, the excess water should be taken back slowly until the knife's edge appears bright. The water level on the knife is now accurate, and the sectioning process can be started (Figs. 5.73, 5.74, 5.75, and 5.76).

When all the settings are done, press the cut button from the control panel or the computerised screen to activate the automatic

Fig. 5.62 Turning the block 1-degree left side to trim the side

cutting function under motor power. ST sections will start cutting in automatic mode. Three digital window displays can monitor sections, i.e., the number of sections, total advance and advance remaining. The number of sections that have been cut since the last Reset can be seen from the display sections. The quality of sections can be monitored through the UM stereomicroscope. Sections are cut at the set thickness, i.e., 400–500 nm, on the down stroke of the specimen arm. In every cut stroke, the sample is retracted slightly on the upstroke to avoid damaging the back of the knife. The section thickness and the cutting speed can be changed in a down stroke of the specimen arm. Once the section is cut, it will float on the top of the water surface. Sections will show colour interference related to their thickness, and a sample with good parallel edges will create a ribbon of sections on the water's surface.

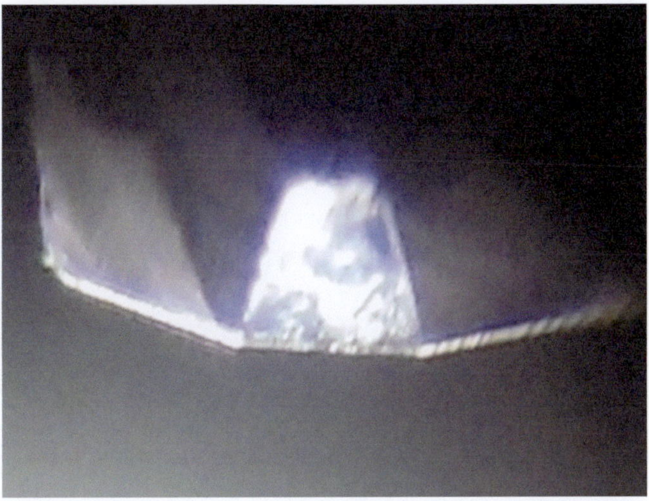

Fig. 5.63 Magnified view of trapezoid face of the block through stereomicroscope of ultramicrotome

Fig. 5.64 Diagrammatic representation of a trapezoid showing two parallel sides from up and down and slopping edge from both sides

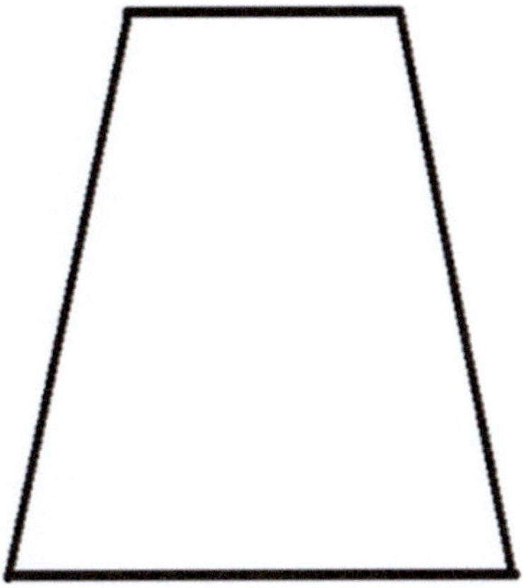

Once the sections are cut enough to pick up, stop the auto-cut and prepare to pick up on a glass slide. Before picking up the ST sections, the glass slide has to be cleaned with a cleaning solution such as alcohol. The sections floated over distilled water are lifted with a thin brush on the glass slide. Sections can be easily trans-

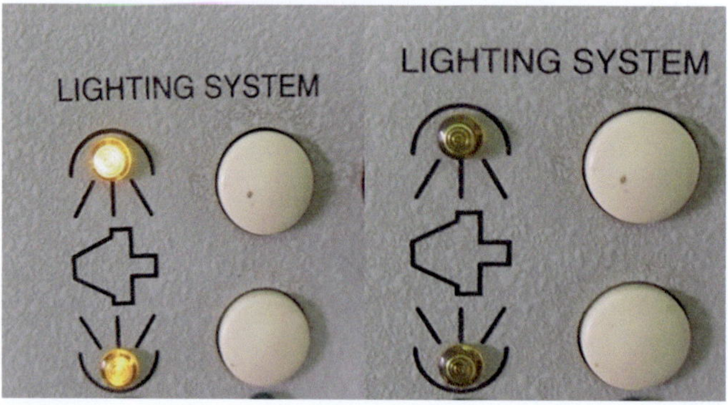

Fig. 5.65 Lighting system on and off mode

Fig. 5.66 Lighting system from computerised control panel

ferred to a glass slide if one drop of distilled water is put on the glass slide prior to picking up sections. The method to pick up the ST sections on the glass slide is,

- The thin zero paint brush should be pushed inside the water and then lifted the sections by it.
- Then, bring the brush to the water droplet of the glass slide and unroll the brush on the water droplet. This way, all the collected sections will transfer and float over the water droplet, and there is less chance of damage and overlapping of sections.

Fig. 5.67 Light from above the sample

Fig. 5.68 Light from below the knife

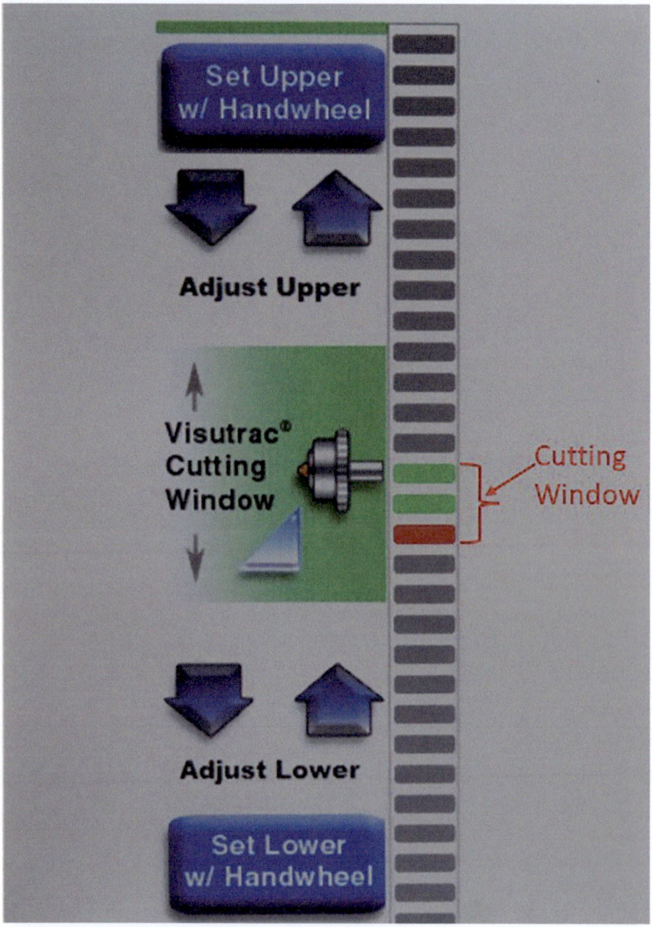

Fig. 5.69 Cutting zone from control panel

Then, the system may reset by pressing the reset button from the control panel. This action will reset the specimen advance mechanism to the beginning of travel and be ready to start a new session (Fig. 5.77).

The glass slide containing ST sections is placed on a multiplate (preheated) and left until the drop of water dries completely. This

Fig. 5.70 Cutting speed from the control panel

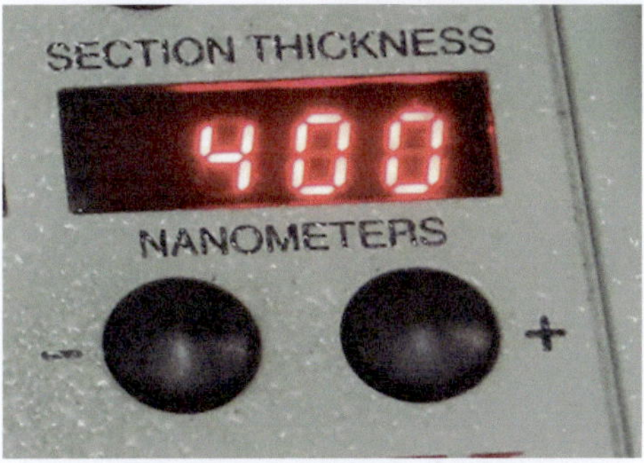

Fig. 5.71 Section thickness for semithin sections from the control panel

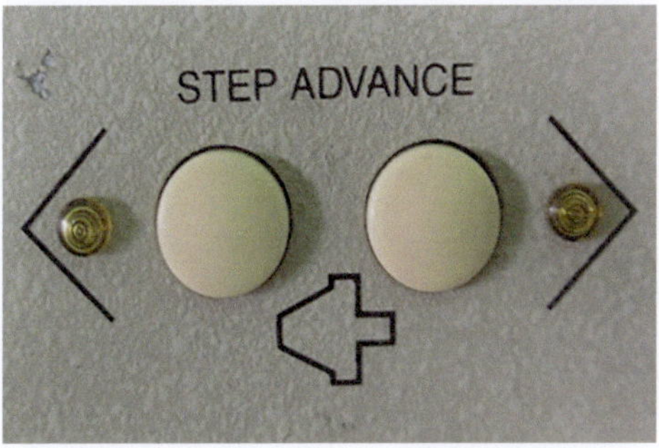

Fig. 5.72 Section advance from the control panel

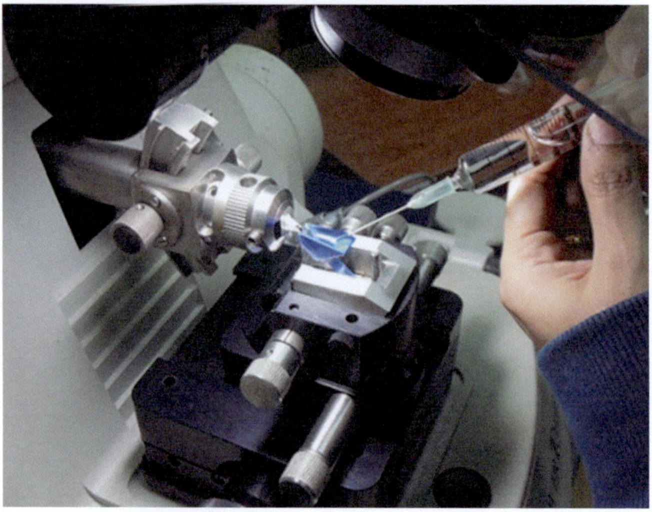

Fig. 5.73 Filling the boat with double distilled water in a convex manner

Fig. 5.74 Bringing the water level to the edge of the knife with an eyelash brush

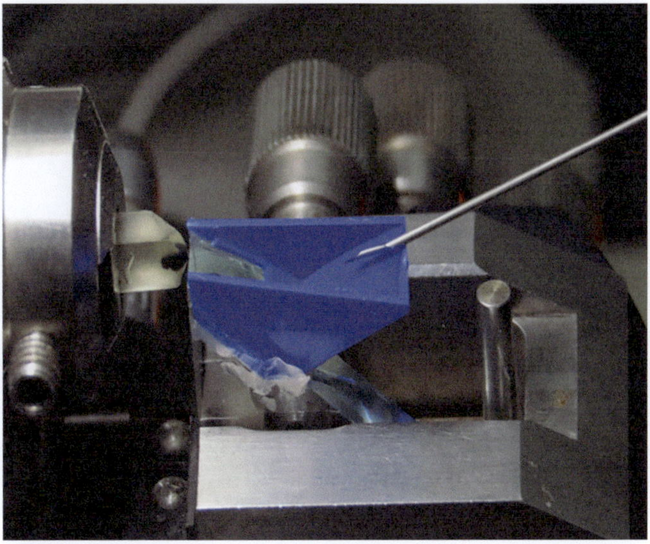

Fig. 5.75 Taking off the excess water by a syringe

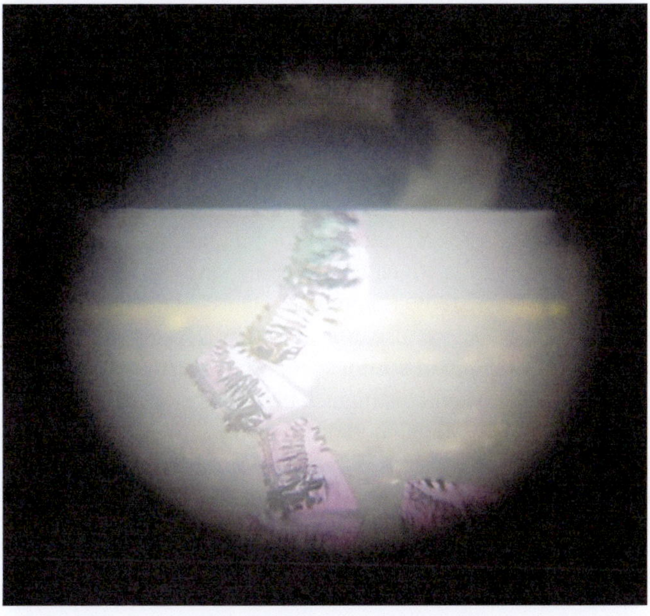

Fig. 5.76 Magnified view of semithin sections through the stereomicroscope

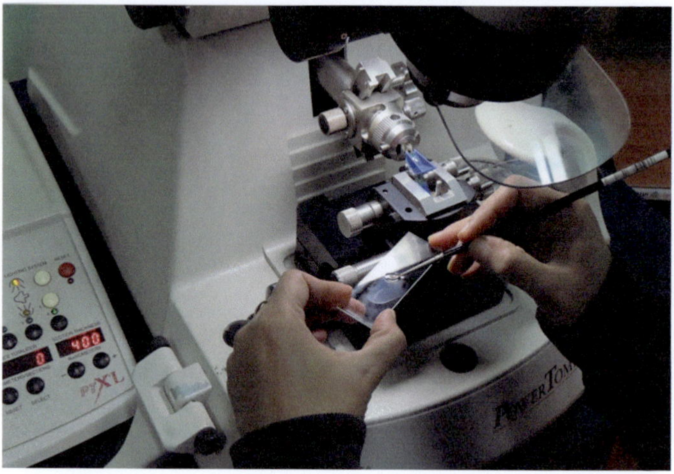

Fig. 5.77 Picking up ST sections onto a glass slide

action will allow the sections to stretch if there is any folding while sectioning and stick the sections to the glass slide (Fig. 5.78).

Staining of ST Sections

Different staining agents have different responses because of their partial miscibility with water. However, in this section, the discussion will only be on toluidine blue stain as this is the most commonly used stain for ST sections for biological samples, which are used as survey sections before processing the UT sectioning step for the TEM study. Some other stains, like methylene blue and periodic acid, are also used to stain ST sections for light microscopy. Staining of sections will provide enough contrast for better visualisation and distinguish the cellular organelles under a light microscope. The method of preparing a toluidine blue stain is mixing 0.25 g of toluidine blue in a 1% sodium borate solution. This stain will allow various elements of tissues and cells to stain differently, yielding sections with a variety of shades of blue (Fig. 5.79).

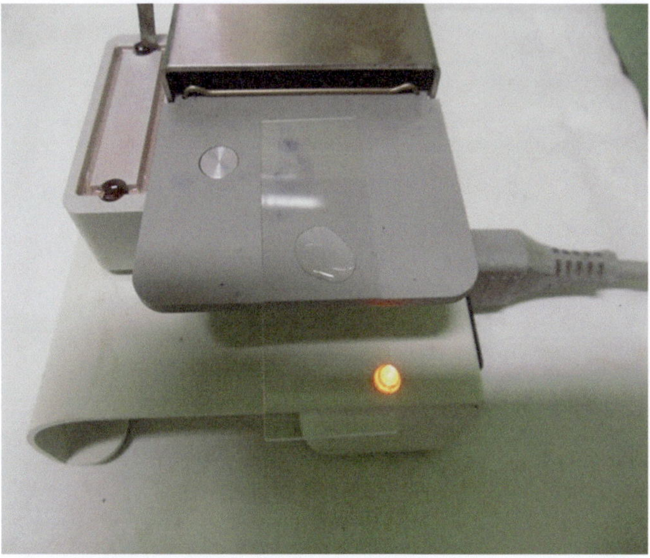

Fig. 5.78 Drying of semithin sections on a multiplate

The method of staining with toluidine blue is that an excess amount of toluidine blue stain should be put over the area of dried ST sections (ensure to cover the entire area containing sections), then leave it a few seconds or a minute until the sides/edges of the stain start drying. Then, the slide may be taken for washing immediately. It is essential to rinse the stain droplet off; otherwise, if the stain gets dried over the sections, nothing can be viewed. This is done by washing the slide thoroughly under running water first and then with distilled water. At this point, the sections will be visible on the glass slide, appearing in blue instead of transparent earlier. It means various elements of tissues and cells are now stained differently with a variety of shades of blue. The glass slide should dry, and a coverslip with DPX should be used for (1) histological study purposes and (2) to check the status of the specimen/tissue under the optical microscope as a survey section before UT sectioning (Figs. 5.80, 5.81, 5.82, and 5.83).

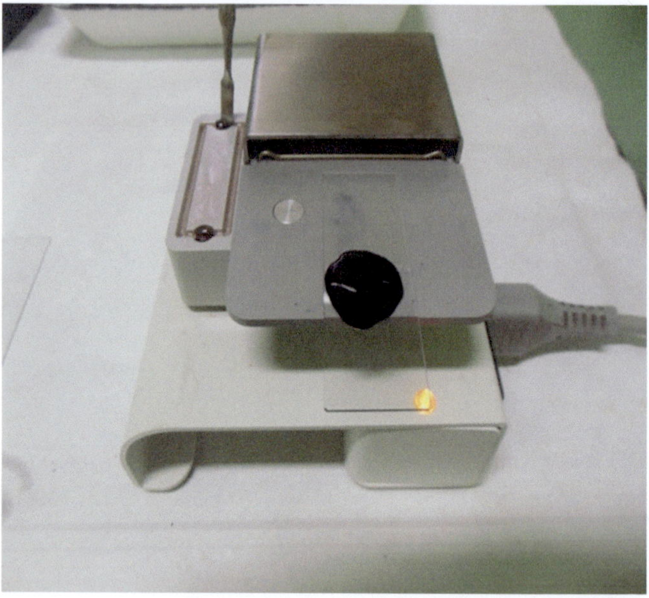

Fig. 5.79 Staining of semithin sections with toluidine blue on a multiplate

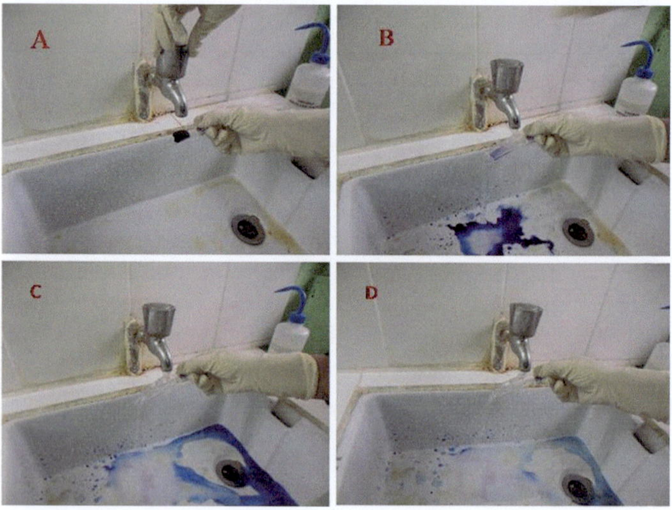

Fig. 5.80 Washing of glass slide under running water (*Steps: A → B → C → D*)

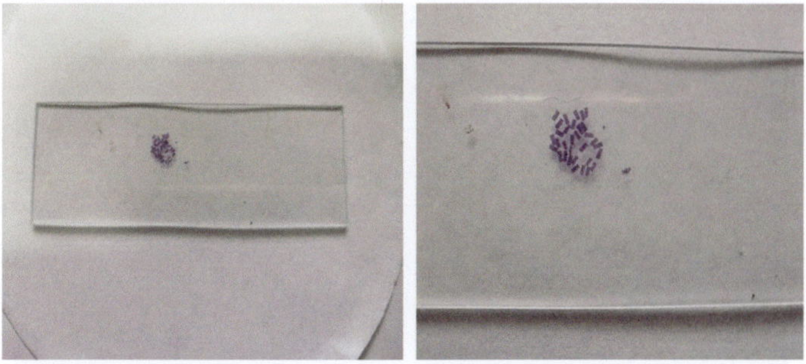

Fig. 5.81 Slides of semithin sections stained with toluidine blue

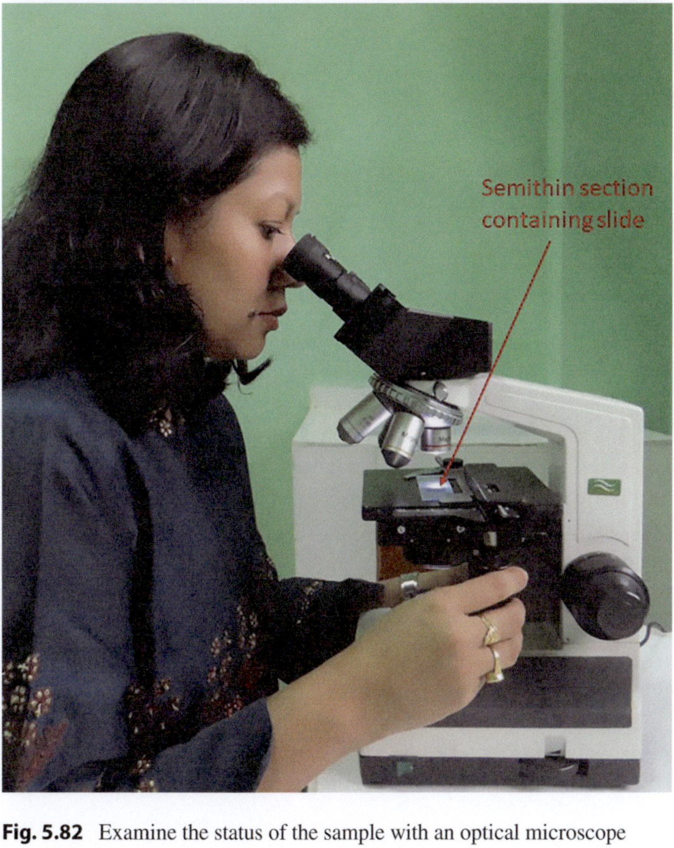

Fig. 5.82 Examine the status of the sample with an optical microscope

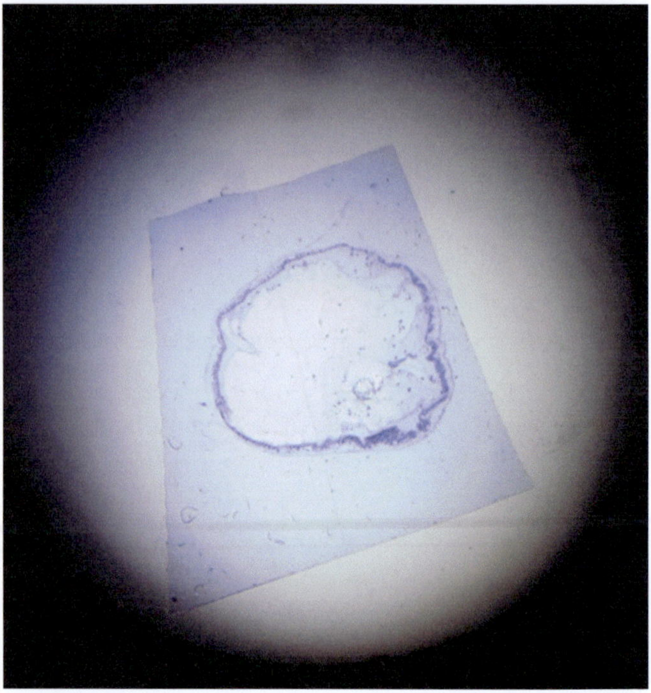

Fig. 5.83 Low magnification semithin section, showing the whole specimen area under a light microscope

This will enable us to know whether any holes are present (not supposed to be present) due to incomplete fixation, dehydration, and infiltration during tissue processing before embedding or any chatter because of the softness of the specimen due to improper heating or less heat while embedding (in this case, the block has to be reheated for a longer time with less heat according to the need of the block status). At this stage, the status of the block will be understood, including whether we need to reprocess or reheat the block or further process it for UT sectioning. If everything is fine, locate the area of interest within the ST section for final trimming prior to UT sectioning. It is worthwhile to mention here that (i) the block has to reprocess if excess/lot of holes are present, (ii)

Fig. 5.84 A TEM micrograph showing the presence of big holes in low magnification

reheat if the specimen area or the whole block is soft, (iii) the block may proceed further to cut UT sections if the fine structure/organelles in the sections are observed under an LM.

Sometimes, some small holes are not visible under an LM, or if the holes are ignored and the same block is further processed for UT sectioning, it may later appear while viewing the sample under TEM, which is of no use (Figs. 5.84 and 5.85).

Final Trimming of Block Face for UT Sectioning

For the final trimming of the block face, the same technique has to be followed as in the case of steps *for machine trimming*. The only

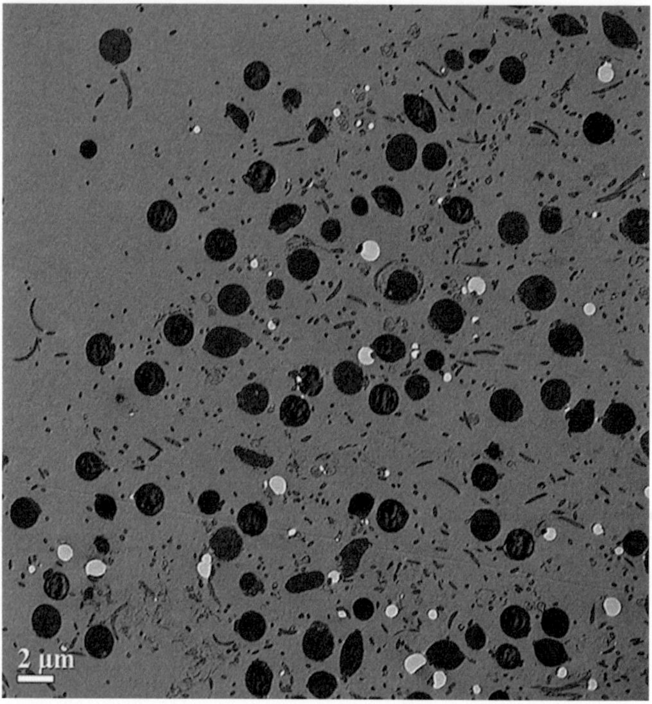

Fig. 5.85 A TEM micrograph showing the presence of holes in high magnification

difference here is that the block face must be very small as the UT sections are very thin (60–90 nm) in nature. If the face of the block is large, like in the case of ST, while cutting the UT sections, the sections may crumple. Another reason is that the knife can induce periodic thickness vibrations in the cut that can damage the knife blade itself. Since the area of the knife edge occupied by the large section will be wider, there are high risks of damaging the knife's edge or the life of the knife soon. So, after examining the ST sections under an LM, only the required area of the sections should be kept, and the remaining area should be trimmed off to get a small face of the block.

It is to be mentioned here that the face of the block was trimmed to trapezoid for ST sectioning for this reason so that the area of study/interest, selected within the ST section under LM examination, can be located/found again on the trapezoid face of the block, which is still kept in the original position of cutting arm without changing the alignment. In final trimming, the unwanted area from the ST section should be trimmed off, and only the necessary area or area of study/interest should be kept back for UT sectioning. Regarding square and rectangular block faces, it is challenging to locate/find the same area from the ST section and block face due to (i) its shape and (ii) the ST sections will sit randomly on the glass slide, making it difficult to understand the exacting sides of the square and rectangle. The square and rectangular shape of the block face may be helpful if only the ST and UT sections are done separately for separate research work. The preparation prior to UT sectioning and the procedure are mentioned below.

A freshly prepared glass knife attached to the boat or a diamond knife can be used for cutting UT sections. Features and preparation of glass knives are discussed earlier in this chapter. Glass knives are economical compared to diamond knives. However, a diamond knife can produce excellent and thin UT sections up to 5 nm, provided extreme care is taken while handling the knife. In connection to this, the diamond knife's features, handling, and care are summarised below. The diamond knife's use and cleaning in ultramicrotomy is also documented here.

Features of Diamond Knife for Ultramicrotomy

The diamond knife is one of the most durable yet fragile tools. The diamond edge is exceptionally durable considering the wide range of materials cut successfully from soft to hard. With proper care and handling, the diamond knife can withstand a great deal of use.

Handling and Care

- Never allow the knife's edge to contact any other object.
- Accidentally touching the knife's edge with fingers, tools or clothing will nick/spoil the knife.

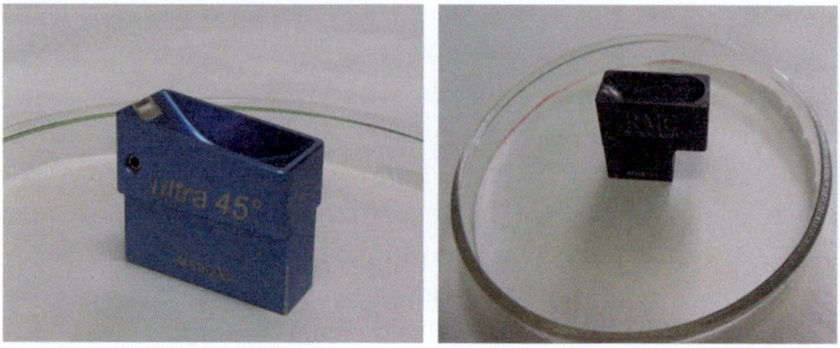

Fig. 5.86 Diamond knives

- The slightest pressure at right angles to the cutting edge, however, can chip it (Fig. 5.86).

Clearance angle: A microtome should be set at 4° as a starting point for diamond knives. If the clearance angle is too high, it will cause "bunching" or "compression" of the specimen block and increase pressure on the cutting edge. Instead, decreasing the clearance angle by one or two degrees can help eliminate this problem. If changing the clearance angle does not eliminate the compression, then the block processing or sample preparation techniques should be evacuated since "soft" blocks can also cause compression. Although the presence of "chatter" depends upon several factors, the clearance angle is the most critical among them. As the optimal angle setting varies from microtome to microtome, the difference setting of clearance angle from 2° to 6° may be tried when chatter is evident. If this technique does not solve it, then the problem might be with the specimen mount or diamond knife holder or may be from a block face that is too large.

Cutting Speed: The cutting speed for the block and the knife is maintained between 0.5 and 2 mm/s depending upon the consistency of the block. Slower speeds can be used to reduce chatter and faster speeds to reduce curling. The knife's edge should be prevented from damaging by maintaining the knife and the block firmly sitting in their respective position (Figs. 5.87 and 5.88).

Fig. 5.87 Clearance angle set at 4°

Fig. 5.88 Cutting speed is set at 1.5 mm/s for a medium consistency block.

Cutting Technique/Procedure for UT Sectioning

- The room where UT sectioning is going on has to be vibration-free, and the air conditioner and dehumidifier, which create vibration, must be off before the operation starts. The door of the room has to be closed. There is even a transparent shield

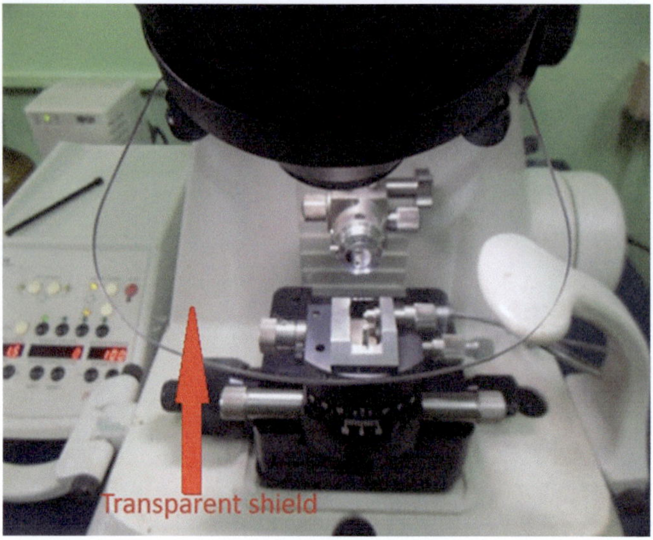

Fig. 5.89 Transparent shield between the microtomist and the cutting area

between the microtomist and the cutting area so that while the microtomist breathes, small vibrations should not affect the section as it is very thin (fragile) or UT in nature. Electronic gadgets like mobile phones should not be kept near the microtome as it might create sudden vibration.

- The stereo microscope of the ultramicrotome is used to visually control all operations (Figs. 5.89 and 5.90).
- Before cutting sections, it is important to align the specimen block and knife carefully onto the ultramicrotome, which consists of a support carrying the knife stage and a sample/block holder that is installed on a vibration-proof platform. Alignment of the block and knife is important so that the first section will cut the entire specimen face.
- Both the specimen block and knife have adjustable orientations. The specimen block is mounted on an arc segment mount, which is a vertically moving arm that moves regularly against the knife edge and then locks it. The device is equipped

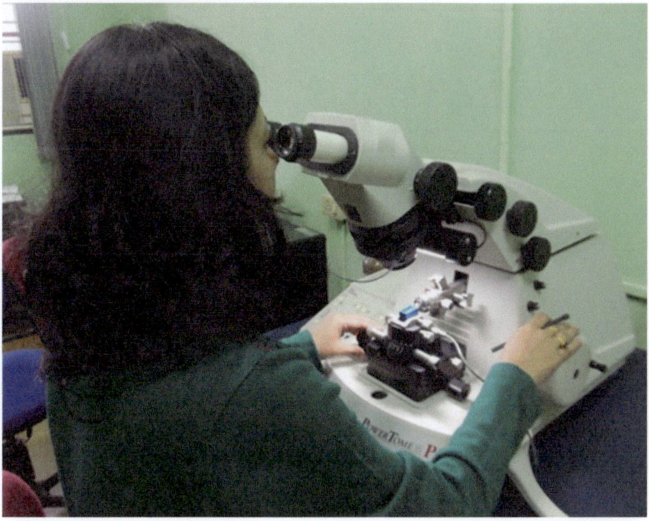

Fig. 5.90 Utilising the Stereo microscope to visually monitor and control all its operations

with a micro to nanometre scale movement system controlling object position. Now, insert a good glass knife with the boat or a diamond knife in the designated place of the knife stage and, position the knife at the height of the indicator, then lock it firmly. Switch on the lights. Adjust the specimen block holder at the same height as the knife edge. Now forward the diamond knife edge towards the block face slowly by using micro feed with the help of coarse adjustment knob. Focus on the specimen face and adjust the orientation arc until the face is shining brightly. At this brightest position, the specimen is aligned to within +1° of the vertical position.

- The face of the specimen reflects the light like a mirror. The reflected image of the knife on the face of the block is visible as a dark shadow across the top of the face. At this position, advance the knife very carefully and slowly towards the block face by using a fine adjustment knob until this reflection can be

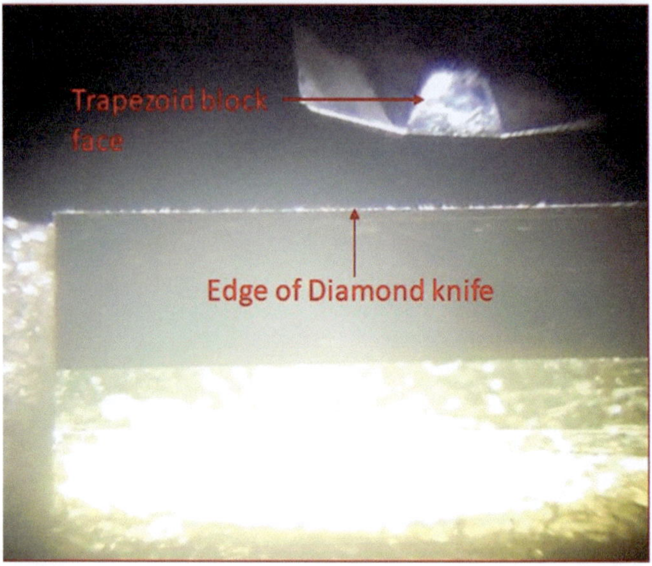

Fig. 5.91 Magnified view of the reflection of light on the face of the specimen

seen. When the silver reflection is a very thin line, that means the block face and the knife are very near, and about to touch, or in another language, we can say that this thin reflection is the real distance between the knife and the specimen. Advance the knife until there is only a narrow silver band between the knife and the knife reflection (Figs. 5.91, 5.92, 5.93, and 5.94).

- All the clamps have to be locked properly. Then, the boat of the diamond knife is filled with distilled water, and the same procedure is followed to fill the boat as in the case of filling the glass knife through, mentioned earlier in ST sectioning for light microscopy from this chapter. The diamond knife edge has to remain wet throughout the ultrathin sectioning process.
- Set the cutting speed between 0.5 and 2 mm/s and nominal feed to higher, like 90–100 nm, as while cutting sections, we want to cut a few thick sections at first and slowly reduce the

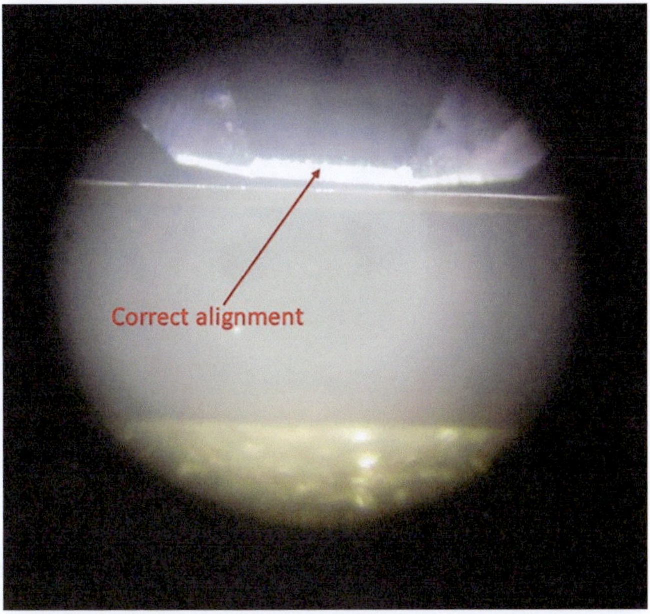

Fig. 5.92 Magnified view of silver refection of light appearing as a thin line on the face of the block on a correct alignment

feed up to 60 nm or until we see silver colour sections starts coming.

- Now, switch on the automatic mode from the manual or computerised control panel and observe under the stereomicroscope until the first section starts. At this stage, if the section delays in cutting, the microtomist can advance the knife by using fine advance knob very slowly or 1.0 μm at once (1.0 μm is one graduation mark on the fine advance knob) till it achieves the first section.

As the specimen starts cutting on UM in automatic mode, the UT sections are cut serially and adhered from one section to the next, forming a ribbon of sections that floats onto the surface of

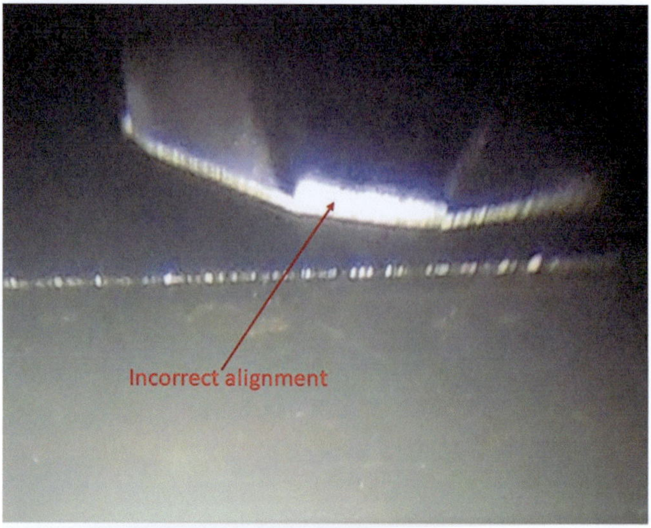

Fig. 5.93 Magnified view of silver refection of light appearing as a thin line on the face of the block on an incorrect alignment

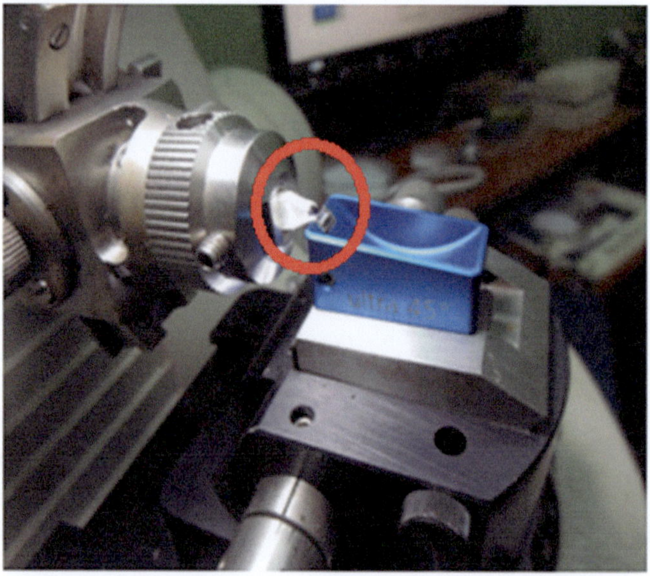

Fig. 5.94 Positioned the block and the diamond knife

the water in the boat of the knife. At this point, the light reflecting off from the water's surface through the sections gives sections a colour and so generates colour interference. The interference of colour corresponds to the thickness of the section, enabling the microtomist to know the rough thickness of the section. The colour of the sections floating on the water's surface ranges from Grey (very thin, below 60 nm) to yellow (very thick, above 280–320 nm). The silver-coloured sections (60–90 nm) have the appropriate thickness for imaging in TEM, and the thicker sections can be used for light microscopy work. Sections with different thicknesses are generated by adjusting the thickness and speed settings of the ultramicrotome. The optimal colour interference chart is given below,

- Grey <60 nm
- Silver 60–90 nm
- Gold 90–150 nm
- Purple 150–190 nm
- Blue 190–240 nm
- Green 240–280 nm
- Yellow 280–320 nm

Generally, in the process of cutting UT sections, allow to cut a few sections a little thicker (3–6 sections) at first and then gradually decrease the section thickness by pressing the (-) button from the section thickness of the control panel until we receive silver colour sections. Now continue the sectioning process and, if needed, change the nominal feed setting to obtain the required section thickness or retain the same silver colour sections. While UT sectioning is going on, the vibration-free environment should be maintained to protect the sections as it is fragile in nature and tends to crumple or spoil with a small mistake (Figs. 5.95, 5.96, 5.97, 5.98, 5.99, 5.100, and 5.101).

TEM images are highly dependent on the quality of UT sections. A good UT section will give a perfect image of the specimen. A small mistake affects the image quality. Some common errors should be avoided during cutting procedures, i.e.,

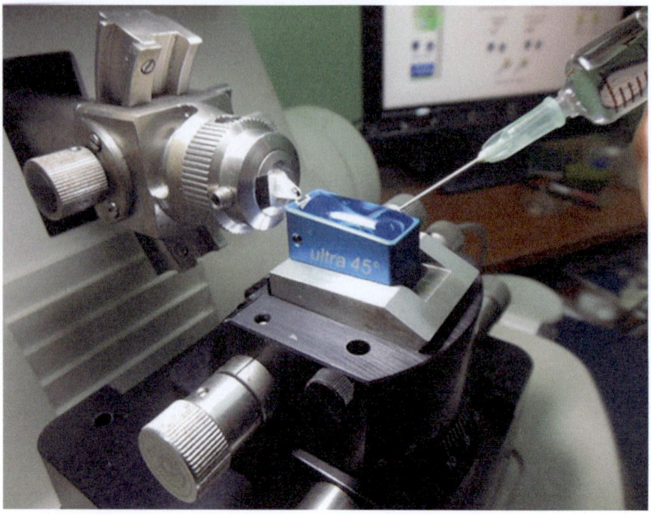

Fig. 5.95 Filling the boat with distilled water in a convex manner

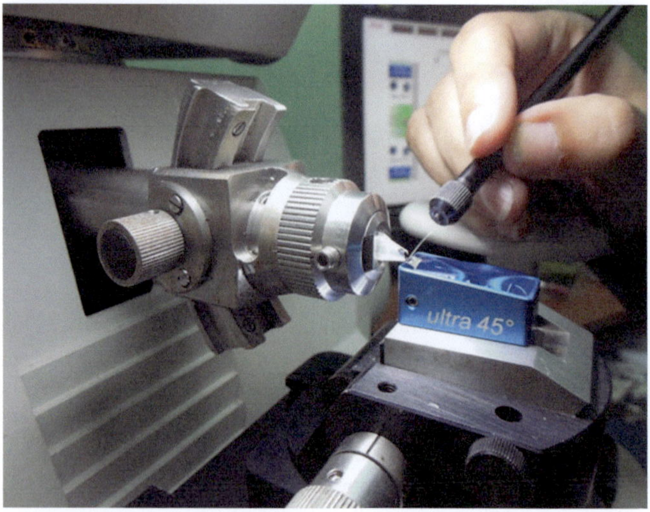

Fig. 5.96 Bringing the water level towards the knife edge with an eyelash brush

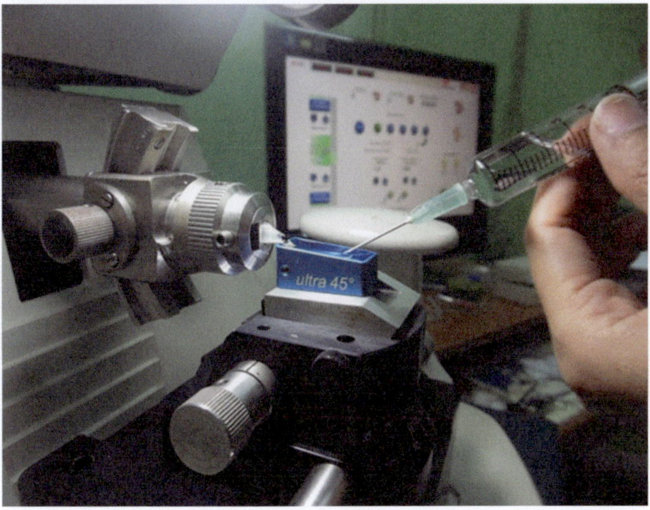

Fig. 5.97 Taking off the excess water from the boat

Fig. 5.98 Reference colour scheme for section thickness (image curtesy, RMC)

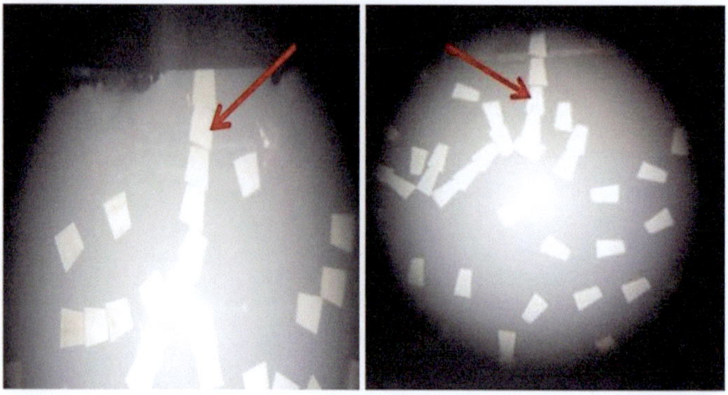

Fig. 5.99 Serial ultrathin sections of silver colour floating on the surface of distilled water

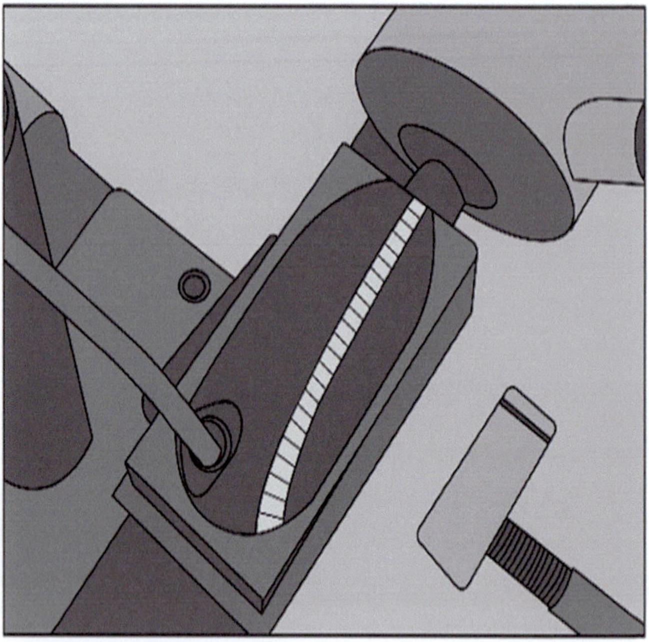

Fig. 5.100 Diagrammatic representation of cutting of serial sections

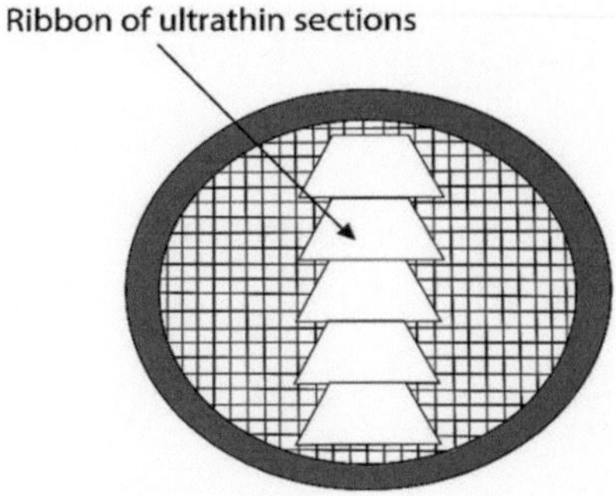

Fig. 5.101 Diagrammatic representation of ribbon of ultrathin sections on a grid

Knife marks: Knife marks are cut marks or tears, generally with uneven spacing and width parallel to the direction of sectioning. The reason for knife marks is due to any debris or bluntness on the edge of the knife. Sometimes, it is caused by the presence of artefacts in the specimen block. So, to get rid of knife marks, ST sections should always be cut before UT sectioning to know the status of the block. Also, use a clean and sharp knife for UT sectioning.

Thick sections: If the UT sections are cut thicker than the usual 60–90 nm thickness, which corresponds to silver colour interference or cannot achieve ultrathin sections because of block consistency due to an error in processing or for any other reason, the thick section will not give a clear vision of ultrastructure of the specimen under TEM.

Chatter or Compression: Chatter or compression are stress marks only viewed under TEM. It is caused by vibration in the room or from a soft block. These problems, as mentioned above, are briefly described in Chap. 6, and some reference micrographs are presented below (Figs. 5.102, 5.103, 5.104, 5.105, and 5.106).

- When sufficient sections have been cut and accumulated enough on the top of distilled water, stop the arm on a down stroke by pressing the cut button from the manual or computerised control panel immediately after a section has been cut (Fig. 5.107).
- With a fine eyelash brush, the thicker sections (initially cut in the beginning or sometime in between) can be pulled back-

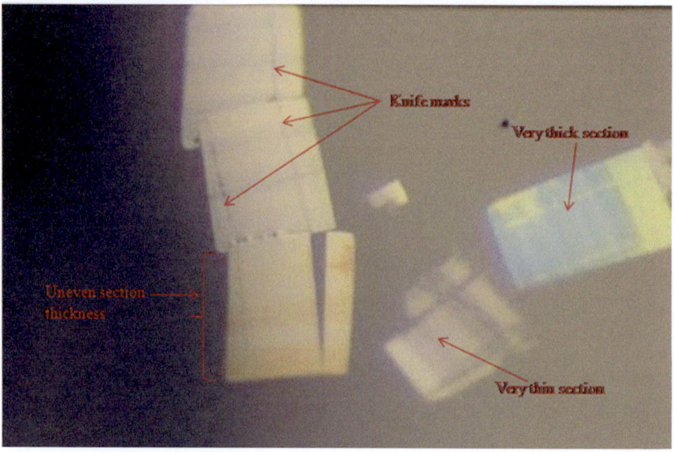

Fig. 5.102 The above picture shows knife marks, uneven section thickness, thick and thin sections in one frame. (Image courtesy to Late Prof. Sudip Dey)

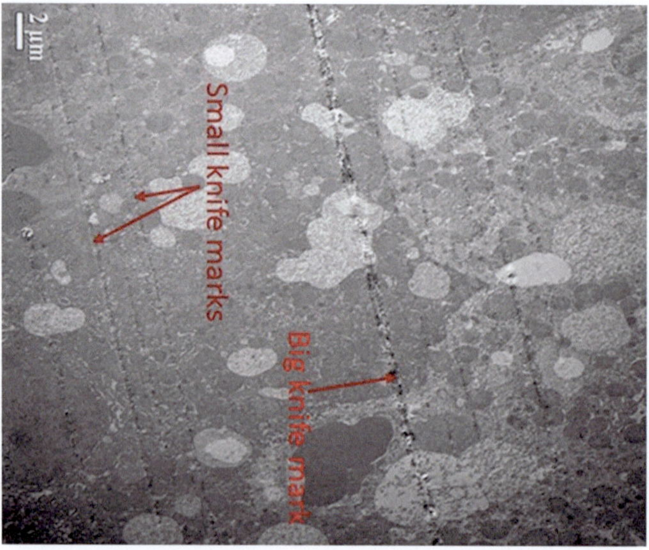

Fig. 5.103 A TEM micrograph with big and small knife marks

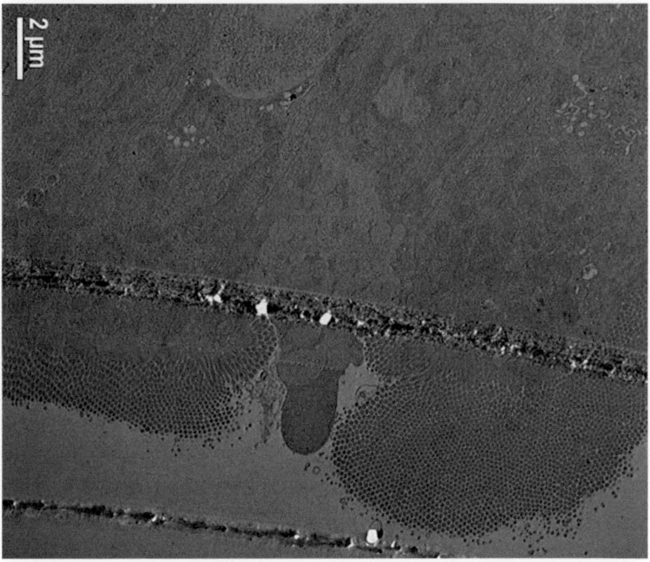

Fig. 5.104 TEM micrograph shows big knife marks

wards and removed entirely from the boat, leaving only the thin silver colour sections.

– Sections are compressed during the cutting process. The sections are now exposed to chloroform vapours to reduce compression by stretching them. This action is done by dipping a small piece of filter paper into chloroform and then gently moving the paper above the sections carefully so that the paper does not touch the sections and see that chloroform does not drop down to the sections/water as it will spoil the sections and may cause harm to the knife. This technique will stretch the sections if there are tiny foldings while sectioning. The sections are now ready for picking up. The selected UT sections can be collected onto a grid (about grids will be discussed later in this chapter). Grids made up of copper (Cu) are commonly used. Other than Cu grids, gold, platinum, nickel, and stainless steel are also available and are used for specific purposes (Figs. 5.108, 5.109 and 5.110).

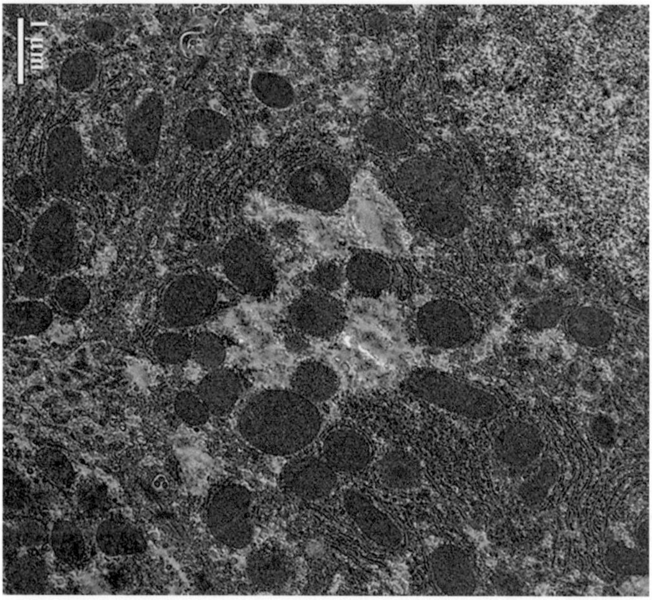

Fig. 5.105 TEM micrograph of thick sections where the cell organelles are not clearly visible

Pick Up the UT Sections onto a Grid

The sections floating on the water are lifted from below with the help of a tweezer. For the collection of sections, select a proper grid (meshed or holed and/or mesh size, etc.) for the specific sample that will support the sections. Free the ribbon from the knife's edge or bring the sections backwards by pulling it with the help of an eyelash brush or single hair brush. Hold the grid with a cross tweezer (forceps can also be used) and lower it vertically into the water trough until it reaches under the water surface. With the other hand (left), guide the sections with the eyelash brush towards the grid held on their right hand and let the first section touch the grid. Slowly remove it from the water at the grid's vertical position. By this technique, sections will be pulled out and unfolded along with the grid, and no sections will overlap. The wet grid-containing sections are now dried by keeping it on to a filter paper containing a Petri dish.

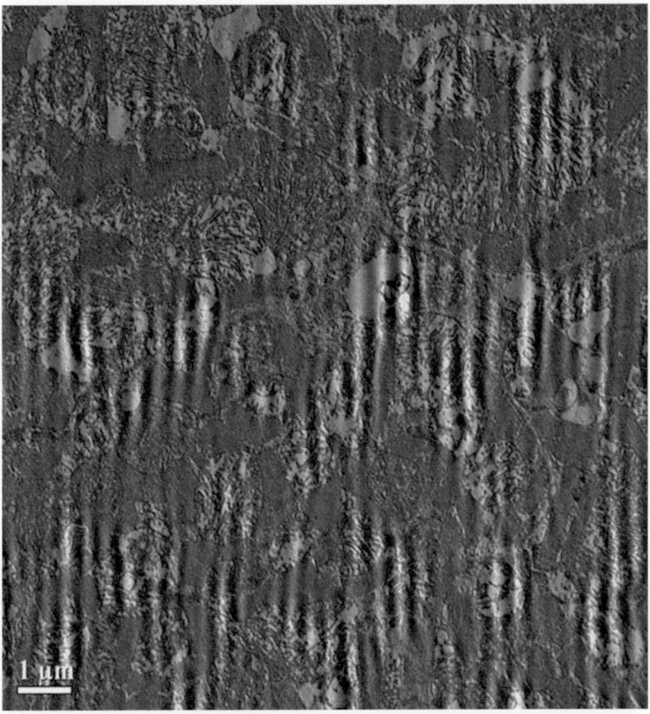

Fig. 5.106 TEM micrograph showing chatter

If the sections are scattered, picking up more sections on a grid becomes difficult. In such cases, the liquid from the boat of the knife can be slightly removed with the help of a piece of filter paper from the opposite end of the knife edge. This process will bring the sections close to each other and allow picking the section up to be easy (Figs. 5.111, 5.112, 5.113, 5.114, 5.115, 5.116, 5.117, 5.118, and 5.119).

- Now loosen the knife clamping knob and manually pull the knife stage backwards carefully so that the knife edge is not touching anywhere; remove the knife from the knife holder, then take it for washing.

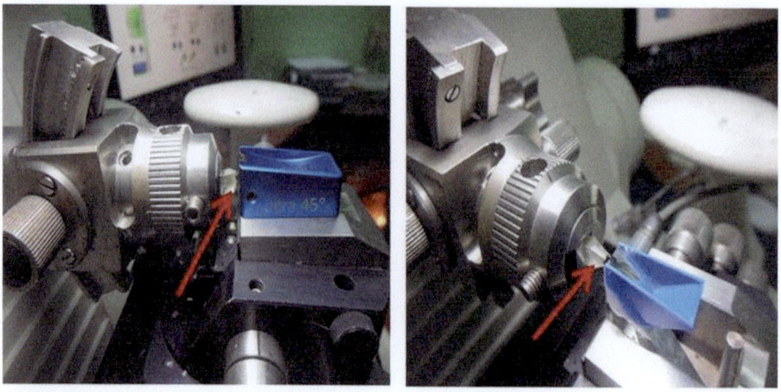

Fig. 5.107 Specimen block arm is in stop position

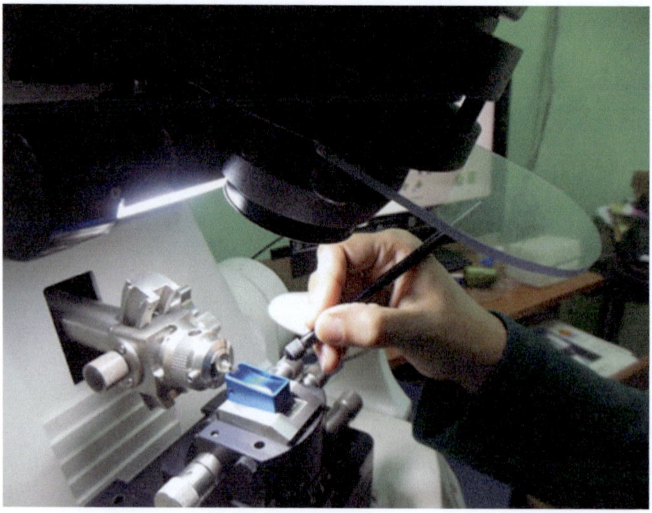

Fig. 5.108 Pulling back the thicker sections

Washing of Diamond Knife

Throw the water-containing sections with force into the wash basin so that no sections will stick to the knife edge; flash the knife edge and the boat thoroughly with double distilled water to ensure no sections are sticking to the knife and the edge.

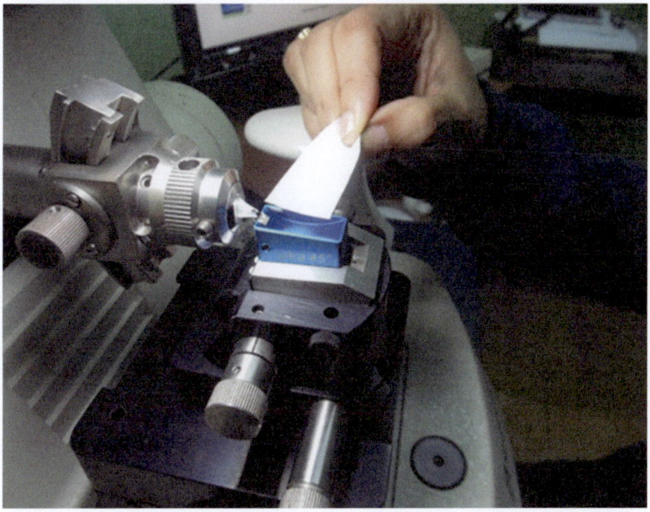

Fig. 5.109 Sections are stretched by exposing them to chloroform vapours

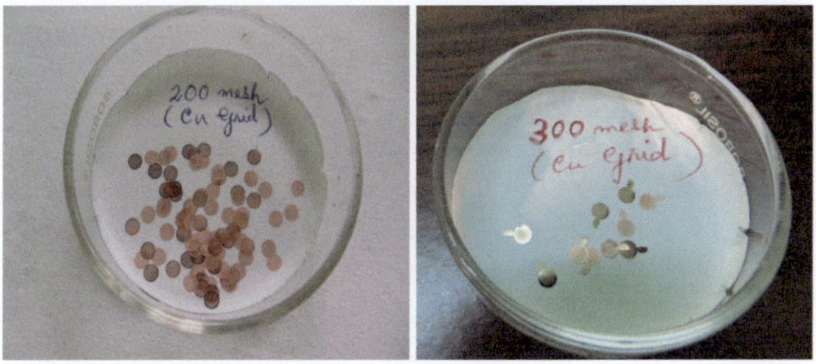

Fig. 5.110 Cu grids of different meshes

- Loosen the clamp of the block holder and remove the block from the holder.

Cleaning of Diamond Knife

It is necessary to clean up the edge of the knife and knife boat every time after every sectioning procedure is completed. Some methods can be followed to clean up a diamond knife given below,

Fig. 5.111 Pushing the grid inside water and placing it under the sections

Method 1: **Routine Cleaning**

Before each microtomy session, an examination of the knife edge is essential. However, at each microtomy/sectioning, extreme care should be exercised to ensure that no sections or section debris gets stuck to the edge of the diamond knife. This practice can be done with an eyelash brush or similar flexible fibre to remove such contamination by carefully pulling the sections backward or teasing away from the knife's edge.

After each sectioning is over, carefully pull back the remaining sections or unpicked sections on the boat without touching the edge of the knife with an eyelash brush, then dump the water from the boat with force. After that, rinse the knife with doubled distilled water with the help of a syringe by flashing each corner of the knife. Next, carefully blot excess water from the boat with a tissue, careful not to touch the knife edge. Alternatively, use a gas duster or clean pressurised air to blow the water off from the knife

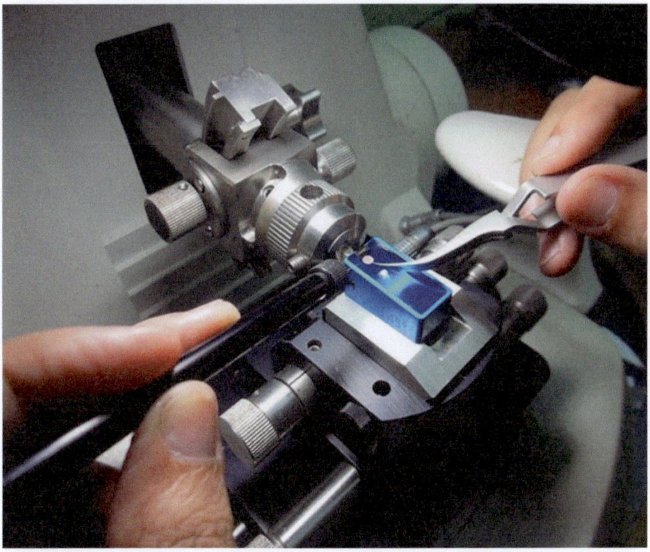

Fig. 5.112 Guiding the sections with the eyelash brush towards the grid held on the right hand and allowing the first section to touch the grid

to dry the knife edge if going for the next microtomy session immediately. Otherwise, the knife can be left for air drying. After drying, carefully place the knife in its box (Figs. 5.120, 5.121, 5.122, 5.123, 5.124, and 5.125).

Method 2: Cleaning with Alcohol/Elaborate Cleaning

After routine cleaning, there is a need to clean the knife edge properly with alcohol. Cleaning diamond knives regularly will give the best results in sectioning/microtomy.

Suppose any sections or debris have dried down or are stuck on the diamond knife's edge, and the knife's edge does not get wet properly while filling the boat with water. In that case, the elaborate cleaning or cleaning procedure with alcohol should be followed. The steps are as follows:

- Take a polystyrol rod bevel and deep the rod into 100% ethyl alcohol, leave it for a few seconds, and then shake the rod to

Fig. 5.113 Picked sections on the grid

remove the excess alcohol. Hold the knife approximately at a 45° angle and pass the rod gently without applying any pressure over the knife's cutting edge. This procedure may repeat until the edge is clean (Figs. 5.126, 5.127, 5.128, 5.129, 5.130, and 5.131).

- Alternatively, the boat of the knife may fill with alcohol, wetting the edge of the knife and keeping it for some time, then dump the alcohol from the knife and flash the edge gently with a syringe with alcohol without exerting any pressure (Figs. 5.132 and 5.133).

Method 3

If the knife edge becomes dirty with stuck dry sections or debris, then,

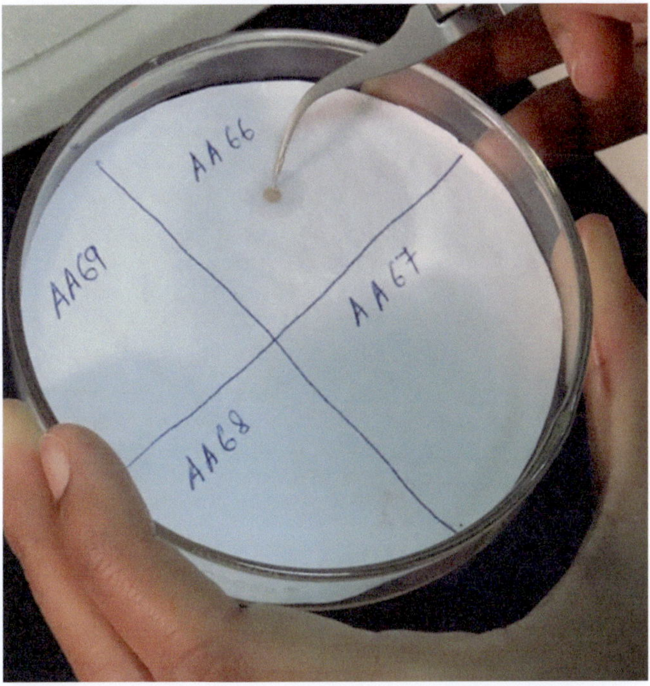

Fig. 5.114 A wet grid-containing sections is dried onto a filter paper

1. Place the knife in a distilled water-filled container safely, then add one or two drops of mild liquid dishwashing detergent to the water-containing knife and let it set overnight.
2. Next day, the knife should be removed from the water and rinsed with cleaned distilled water or proceed to clean the knife using *Method 1* (Figs. 5.134, 5.135, and 5.136).

If the routine cleaning of the knife is followed up after every sectioning, then usually, there is no need for any elaborate cleaning.

Grids

The UT sections are collected on specially-made grids for TEM use. Grids are available as meshed or holed ones and are used

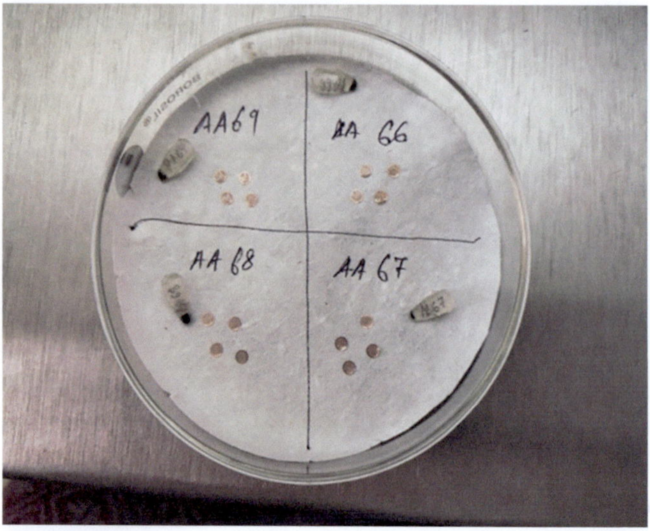

Fig. 5.115 Different specimen blocks with their respective sectioned grids

according to the needs of the specific sample. Grids of 100–500 meshed sizes with different pitches, bar widths, and hole widths are available, and the most commonly used ones are 100–300 meshed for biological samples. Grids are generally coated on one side with specific metals like copper, rhodium, nickel, gold, etc. and are matt on the coated side but bright on the other. The sections are lifted using the matted surfaces of the grid, which allows the sections to stick to the grid face firmly. Generally, all grids are approximately 25 μm in thickness. Grids made of copper are commonly used for biological specimens because of copper's non-ferromagnetic nature, which minimally distorts the magnetic field of the objective lens during TEM operation. The copper grid also rapidly removes heat away and thus prevents thermal expansion and movement of the specimen under electron radiation (Fig. 5.137).

In TEM, a high-energy electron beam bombards and transmits through the sample. Depending on the amount of energy absorbed by the sample, the beam's intensity varies, and an image is made

Fig. 5.116 TEM micrographs with foldings due to error during pick up

where the contrast arises from the beam interacting with the sample. In the case of biological samples, the main components, i.e., carbon, oxygen, nitrogen, and hydrogen, are not very densely present, and the number of electrons they absorb is minimal compared to the intensity of the electron beam. Grids are coated with a film for a better resolution under TEM, which is essential to render support to the sections. Film coating is done on the matted surface of the grid. These films should be extremely thin (20 nm or less) and made of suitable electron-transparent substances, such as plastic or evaporated carbon, without breaking when gently handled or exposed to the electron beam. Coated grids help

Fig. 5.117 TEM micrographs with foldings and precipitates

support particulate specimens like viruses and bacteria; otherwise, without a support film on the grids, the tiny samples might fall through the openings in the grids.

Different Types of Support Film

Different types of film types are prepared to support the grid. Coated grids also helped to minimise astigmatism in the objective lenses and apertures and manage image stability in TEM. Various film types are involved,

Fig. 5.118 TEM micrograph with lots of folding due to an error in picking up

(a) Nitrocellulose: Nitrocellulose is available as strips of material named parlodion, which has to be dissolved in a solvent, amyl acetate before use, and 1–2% solutions of nitrocellulose in amyl acetate are also available as collodion.

(b) Formvar: Formvar or polyvinyl formaldehyde is available as 0.25–0.5% solution in either chloroform or ethylene dichloride (2,2-dichloroethane) or also as powder. To make the solution from powder, put 0.25 gm in a glass container with a stopper seal that is not soluble in the solvent and then add 50 ml of solvent (ethyl dichloride). Now, the solution has to sonicate briefly and then let it set for some time at room temperature. The colour of the solution will be seen as a pale straw liquid after 24 h, and this prepared solution is not light-sensitive and stable for months at room temperature. The solution must be adequately covered with a Teflon seal as it becomes more concentrated because of evaporation.

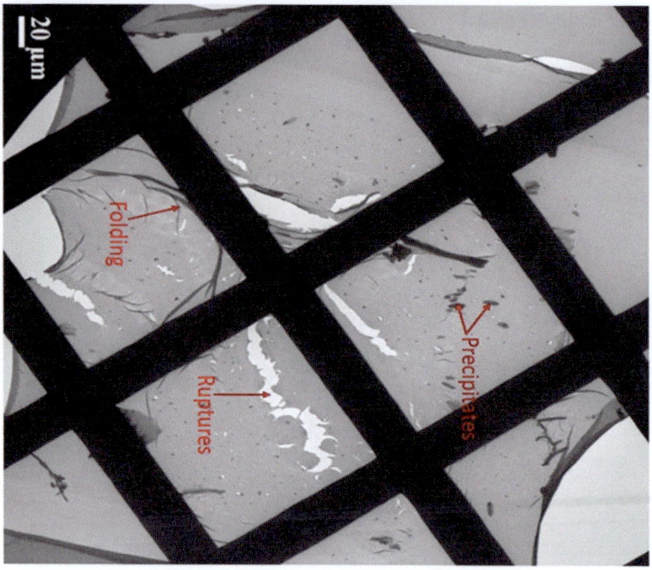

Fig. 5.119 TEM micrograph with precipitates due to staining error, folding due to error while picking up the sections and ruptures due to over-stretching by chloroform

It is worth mentioning here that another polymer, butvar or polyvinyl butyraldehyde, can also be used similarly with formvar.

(c) Carbon: In this method, carbon is evaporated onto the plastic-coated grids (such as collodion, formvar, and butvar). This procedure is done to increase their conductivity under the electron beam and increase their stability.

It is done by placing air-dried filmed grids into a vacuum evaporator where the vacuum has to be about 5×10^{-5} Torr, and then about 10–20 nm of carbon has to evaporate onto the surface of the filmed grid at 45–90°. Now, the carbon-coated grid is coloured light to dark grey.

Method: Coated grids with support film from any of the substances mentioned above are prepared by cleaning the grids first so that the materials will adhere properly to the grid surface.

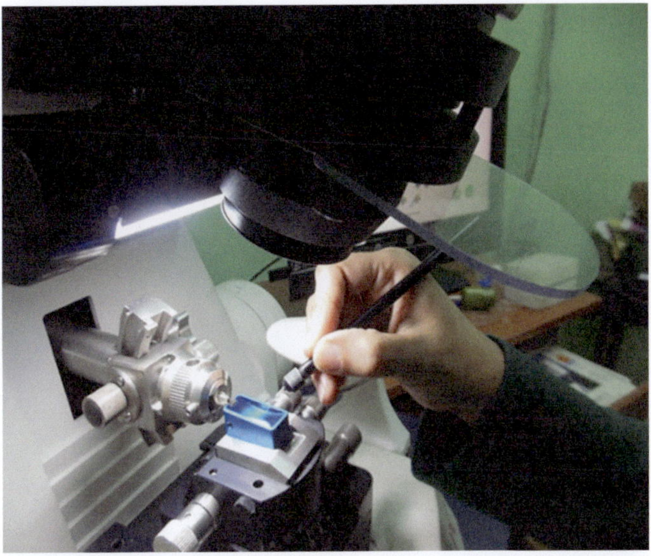

Fig. 5.120 Carefully pull the sections backwards with an eyelash brush

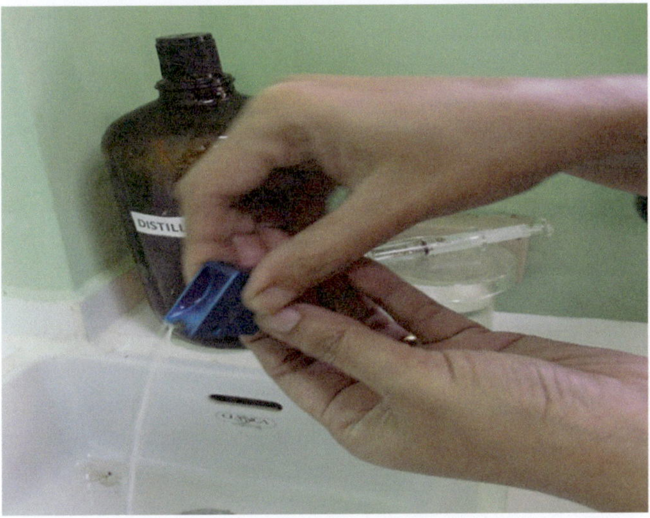

Fig. 5.121 Throwing the water from the knife with force

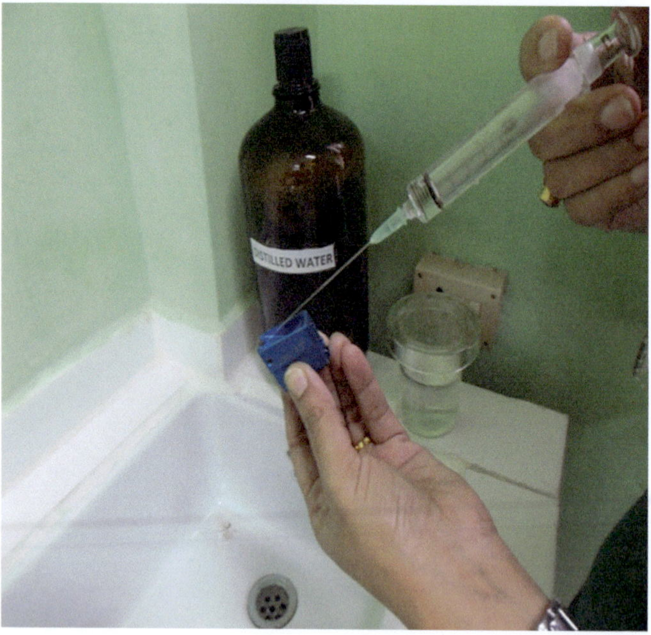

Fig. 5.122 Flashing the knife edge with double distilled water with the help of a syringe

To clean the grids prior to preparation for coating, the following steps are followed:

- The grids are to be put into a BEEM capsule or in a microcentrifuge tube and 0.1 N HCL to it, then the cap and shake for 10 s.
- Pipette out the acid and replace it with 95% ethanol. Now shake it for a few seconds and then remove the ethanol.
- Introduce 100% acetone and shake vigorously, then pipette out the acetone. The pipette tip has to be changed at every step.
- Now empty the BEEM capsule or microcentrifuge tube by shaking or taking the grids onto a clean filter paper in a petri dish and leave it for the acetone to evaporate and dry the grids.

Now, with the help of a forceps or cross tweezer, separate the grids and line them up face down, which means the shiny side

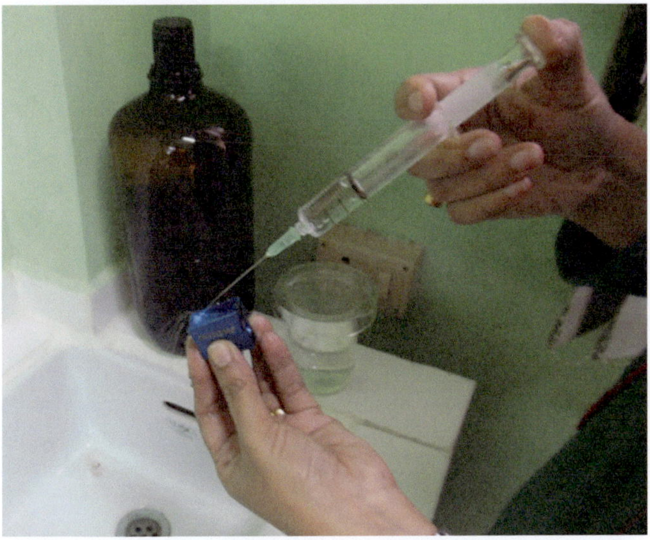

Fig. 5.123 Flashing the boat of the knife

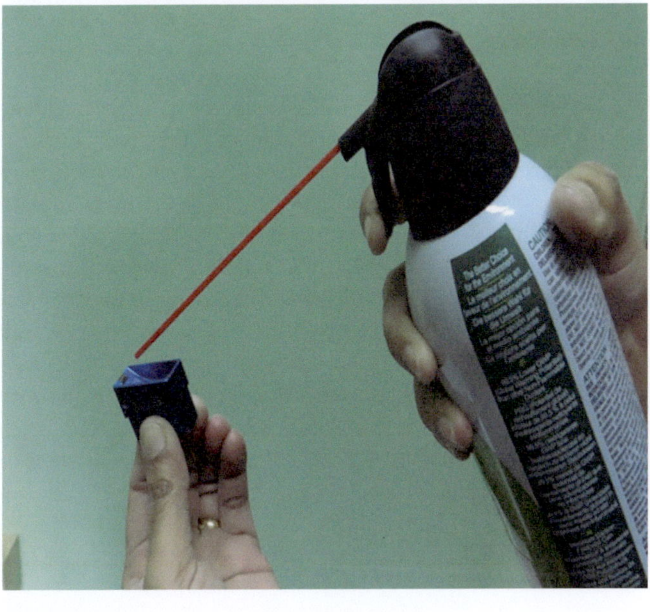

Fig. 5.124 Drying the knife edge with a gas duster

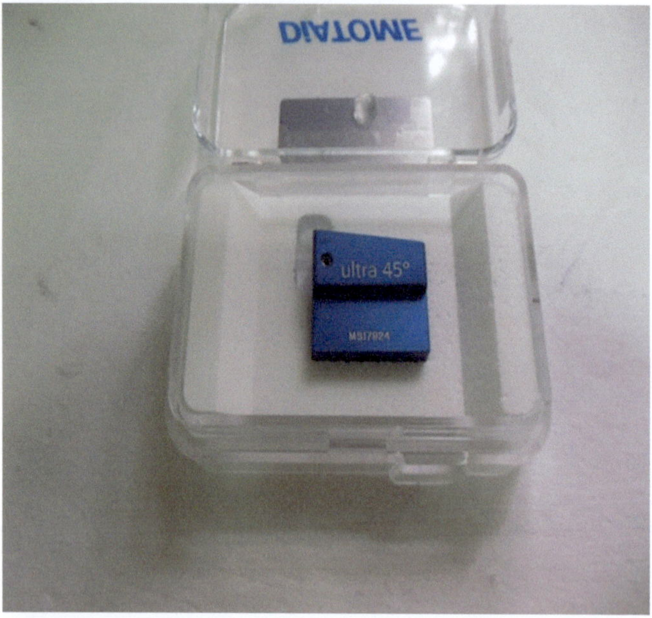

Fig. 5.125 Placing the knife on its box

down on a clean filter paper. This has to be done immediately before grid coating, as they tend to oxidise within a few hours.

In a clean glass vessel, such as a rectangular histological staining dish for racks containing ten slides, keep the grids separate and filled with deionised or distilled water. In the droplet method, 1–3 droplets of a 2% collodion solution are dropped onto the water's surface in the film-casting dish. Collodion is used because it spreads readily on the surface of the water. So, as the solvent evaporates, a plastic film will cover the entire surface of the dish. This film has different thicknesses in different areas, which can be understood from the interference of colours and also wrinkles in places. After the solvent evaporates completely, the grids are to air dry and ready for use.

Benefits of the ultramicrotomy technique:

Fig. 5.126 Polystyrol rod

- This is the only technique directly yielding 30 nm sections of constant thickness.
- It can be applied to a precise sample zone selected with the optical microscope.
- It consists of successive sections of the same specimen and provides large observation areas.

Staining of UT Sections

After the UT sections are picked up onto grids, the sections have to undergo a staining process, and the purpose of staining UT sections is to provide contrast to the biological organelles present in the section while viewing in TEM. Embedding media (with which the specimen is infiltrated and polymerised) and biological materials are chemically similar, which contain large amounts of car-

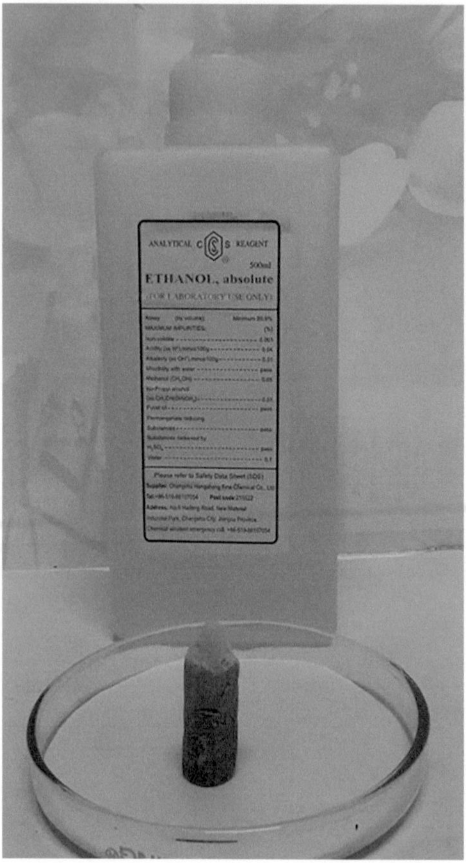

Fig. 5.127 100% ethyl alcohol

bon, hydrogen, oxygen and other fairly electron-transparent atoms, and also a few atoms of large atomic weight. For this reason, major contrast differences between the specimen and the embedding medium do not exist; there is a need to add other materials or introduce stains to develop the contrast. The most effective stains contain heavy metals like lead, uranium, and osmium. These heavy materials have sufficient mass to block or scatter electrons that hit the specimen. If the mass of the stain

Fig. 5.128 Deeping the rod into alcohol

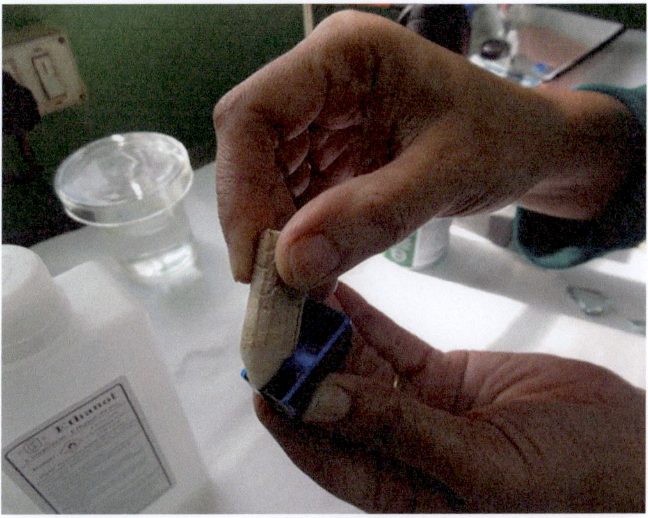

Fig. 5.129 Gently passing the rod over the cutting edge

Fig. 5.130 Throw the alcohol off

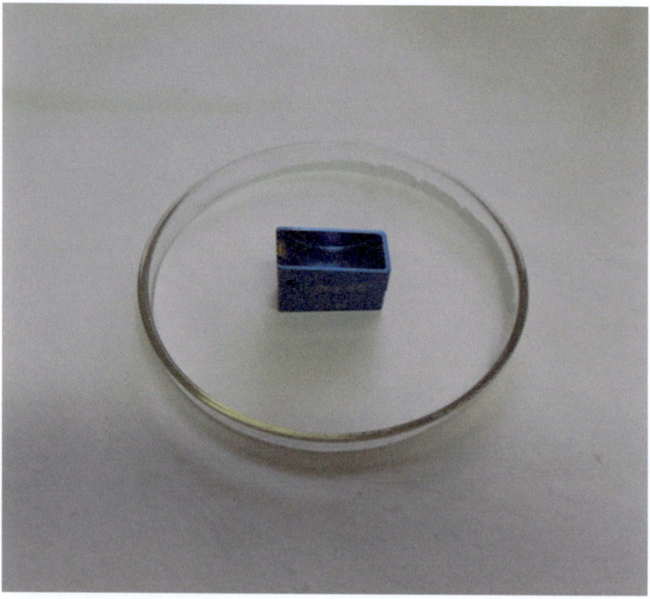

Fig. 5.131 Air dry

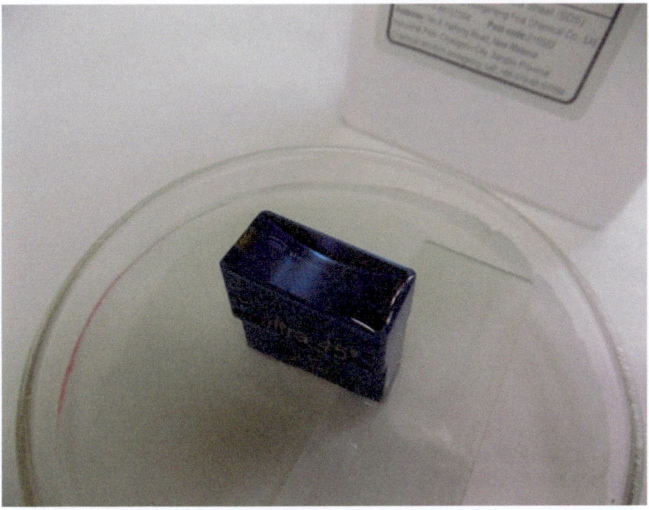

Fig. 5.132 Filling the boat of the Knife with alcohol with the knife edge wet

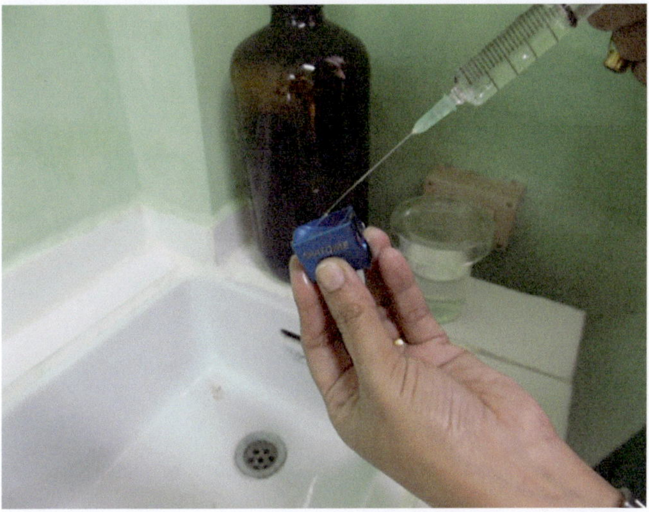

Fig. 5.133 Dumping the alcohol off and flashing the edge with alcohol by a syringe

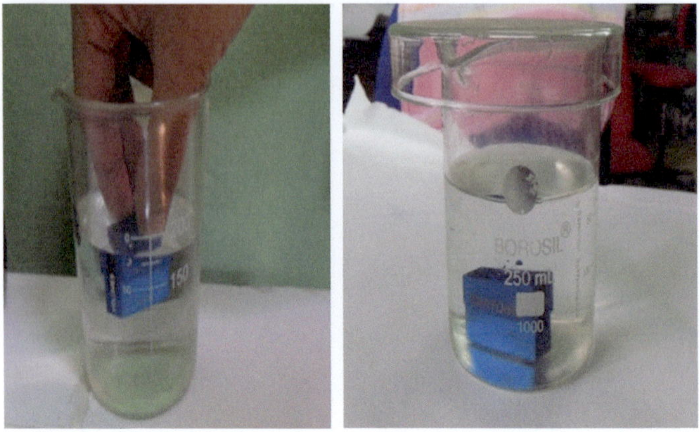

Fig. 5.134 Placing the knife in a distilled water-filled container and add a few drops of mild liquid dishwashing detergent to the water

Fig. 5.135 Rinsing with distilled water

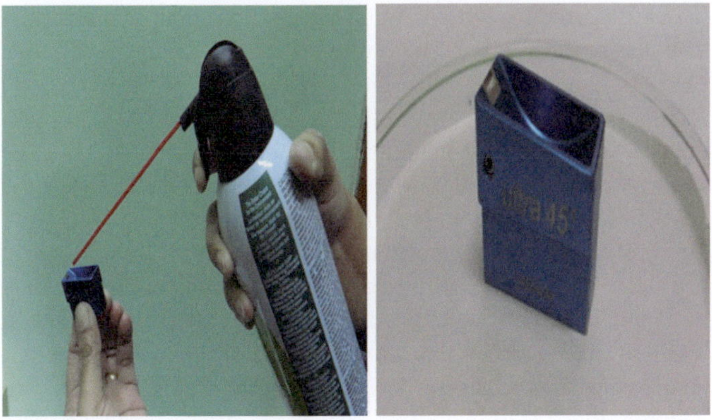

Fig. 5.136 Drying the diamond knife

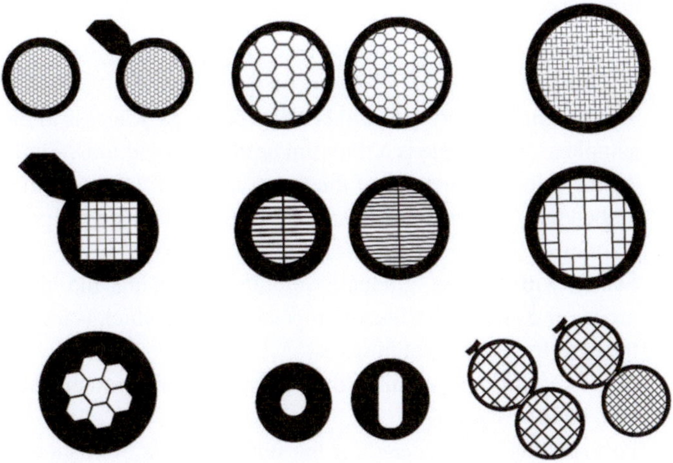

Fig. 5.137 Diagrammatic representation of different types of grids

molecule is larger, it is better to block the beam. To a certain extent, the resolution is also affected by the size of the heavy metal ions deposited during staining. Staining UT sections also helps stabilise specific specimen components during bombardment by the electron beam in TEM.

En Bloc and Post-staining: En Bloc stains are used on wet tissues. Osmium tetroxide, used as a secondary fixative, is considered for lipid fixation and can also be used as an en bloc stain. Another commonly used en bloc stain is uranyl acetate. For this stain, it is used after osmication of sample. The sample has to be rinsed thoroughly with distilled water after osmication as the phosphate and cacodylate buffer used with osmium tetroxide strongly react with uranyl acetate. For staining with uranyl acetate, the washed sample must be placed in 0.5% aqueous uranyl acetate and then kept at 4° overnight. These samples are then rinsed with distilled water, dehydrated, and embedded like the case of regular sample preparation for TEM. The main disadvantage of this procedure is that it is time-consuming, and the effect of uranyl acetate is found more when used as a post-stain (Hayat 1981). Post-staining is done on the ultrathin sections after the embedding of samples is completed.

The advantage of post-staining over en bloc staining is;

- The staining is more uniform because of the easy access to the sample as the sample is a thin film or ultrathin sections,
- As the cellular structures of the tissue are more stabilised after embedding, there is less chance of tissue damage in post-staining,
- Stains will not be gradually eliminated during the post-processing of samples in case of post-staining, which is indeed the case for en bloc staining.
- In the case of post-staining, the contrast develops much faster than en bloc staining.
- Post-staining is considered for the study of quick evaluation of different staining procedures and also for different stains, as the effect can be found on serial sections from the same block rather than numerous samples separately in the case of en bloc staining.

Staining UT sections is necessary for better contrast and better resolution of the cellular structures of the specimen. Although the specimens are stained with osmium tetroxide during the post-fixation step of sample preparation, the additional heavy metal stain is added for good contrast to the sections. Commonly used post stains are,

Uranyl stain: Uranyl stain is used as en bloc, post-stain, or negative stain. Uranium is the heaviest stain molecule, with an atomic number of 92 and a weight of 238. It binds strongly to phosphate groups (nucleic acid and phospholipids) and carboxyl groups (amino acids of proteins). The pH of uranyl stain is usually adjusted to 3.5–4. Different varieties of uranyl stain include uranyl acetate, uranyl formate, uranyl magnesium acetate, and uranyl nitrate. The most commonly used stains are either aqueous or alcoholic uranyl acetate. To prepare the uranyl acetate stain, make a saturated solution by adding excess uranyl acetate to 10 ml of filtered 50% ethanol in a 15-ml centrifuge tube. Shake the solution vigorously for 2 min or more and spin down hard to allow the excess of uranyl acetate to settle down on the bottom of the tube. The solution must be centrifuged now, and the supernatant must be taken for staining. The solution is ready for use.

Method of staining: A drop of uranyl acetate solution is put onto a fresh square of parafilm placed on the base of laminar flow or a petri dish. Now, the grid-containing sections has to emerge in the drop of stain with the section side up. After about 30 min at 30 °C, the grid is removed from the drop of stain and washed immediately with running distilled water thoroughly to remove all the uranyl acetate stain and then dried on filter paper (Figs. 5.138, 5.139, 5.140, and 5.141).

Following up the uranyl acetate stain with lead citrate in the sequence is essential to achieve maximum contrast, as lead will interact more strongly with tissues that have already stained with uranyl acetate.

Lead stains: The lead atom is a very effective electron scatterer with an atomic number of 82 and an atomic weight of 207. In some cases, it is also used as an en bloc stain but mainly as a secondary post-stain along with uranyl acetate. There are a variety of lead stains, such as lead hydroxide, lead acetate, lead tartrate, etc., but it is very tedious because of its formation to lead carbonate extremely fast with air. Lead citrate, introduced by Reynolds (1963), is now a widely used lead stain having a pH of 12–13. The tendency to form lead carbonate by interacting with air is reduced

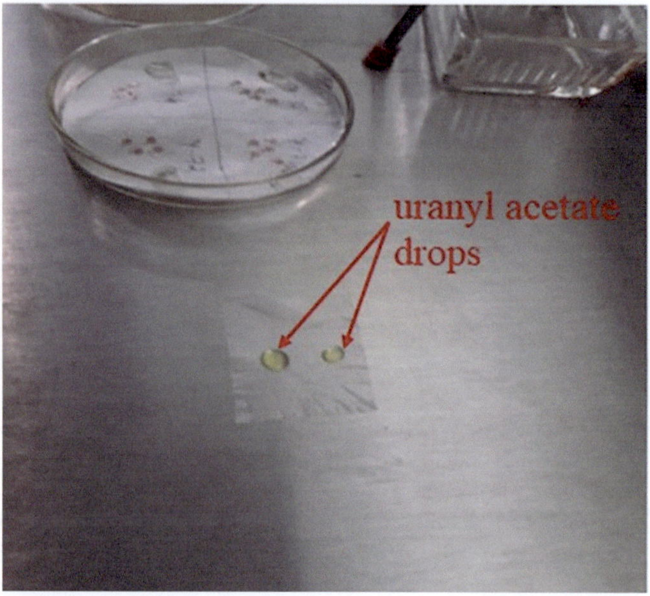

Fig. 5.138 Drops of uranyl acetate stain on a parafilm placed on the base of laminar flow

in lead citrate as lead is strongly chelated with citrate. To prepare a lead stain, add one half of a pellet of sodium hydroxide to 12 ml of double distilled water in a 15 ml centrifuge tube. Shake well to dissolve. Add 50 mg of lead citrate, shake well for two or more minutes, and centrifuge the solution. The solution is ready for use.

Method of staining: Centrifuge the prepared stain, take a drop from the supernatant and put it onto a fresh square of parafilm placed on the bottom of a petri dish. Place the grid carrying the section down onto the stain for 5–10 min at around 30 °C. Lead precipitates quite readily with carbon dioxide (CO_2) and oxygen (O_2) from the air and forms lead carbonate ($PbCO_3$), which may appear as fine or large granular deposits while viewing the section on TEM. This stain is advised to be done in the dark, and a face mask to cover the mouth and nose is used. Also, use carbon dioxide scavengers such as KOH and NaOH to surround stain drops

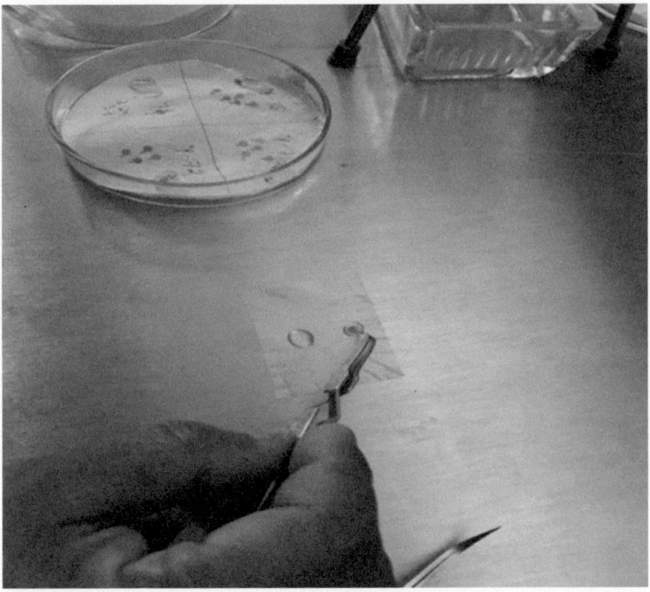

Fig. 5.139 Emerging the grid in the drop of stain with section side up

and follow up the staining procedure inside a laminar flow. Excess stains are washed off by running double distilled water and then drying the grids on filter paper (Figs. 5.142, 5.143, 5.144, and 5.145).

Phosphotungstic acid: Phosphotungstic acid (PTA) was first used by Hall et al. (1945) to impart electron density to biological specimens. This is an en bloc and negative stain for bacteria, viruses, and subcellular particle suspensions. PTA is used as en bloc staining for those samples that are difficult to section.

PTA and Chromic acid: PTA and Chromic acid are used as post stains, introduced by Roland et al. (1972). The composition of PTA and chromic acid stains the plasma membranes of plant cells. This stain is used to analyse the isolated cytomembranes to determine the volume of plasma membranes in their samples.

Barium Permanganate: Permanganates were used as post-stain in the past, but they are of little use due to precipitation on sec-

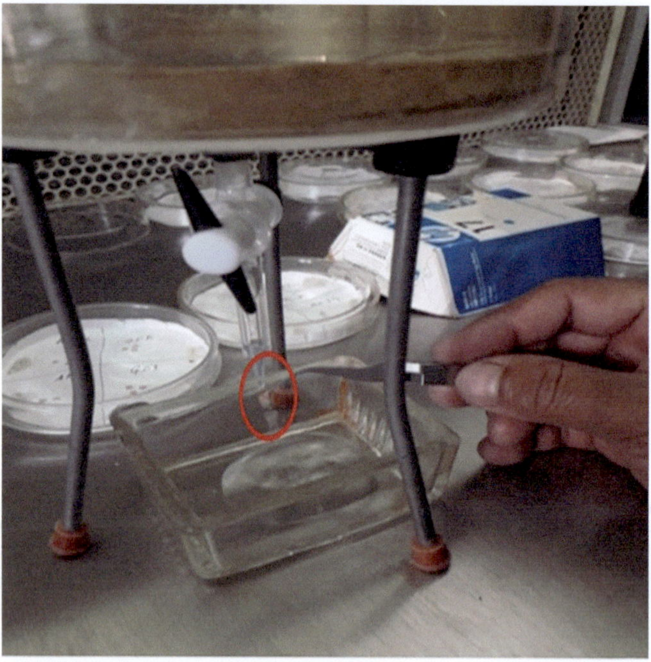

Fig. 5.140 Washing the excess stain off with doubled distilled water

tions. In 1968, Hohl et al. found a way of using barium permanganate as a stain for cellulose fibrils in cellular slime mould, acytostelium. A solution is prepared by adding 0.5–1% in distilled water. This stain is used 1–2 min after lead staining and then washed thoroughly with distilled water.

Microwave Staining: Microwave staining is used mainly for semithin sectioning and ultrathin section staining in some cases. However, this technique has proven faster than the standard double staining method with uranyl acetate and lead citrate. Estrada et al. (1985) developed a procedure of utilising a 400 W, 2450 MHz microwave with a rotating platform to accelerate uranyl acetate and lead staining. In this method, the section containing the grid has to be immersed in 3–4 ml of stain for the microwave step, and the grid has to be microwaved for 15 s in

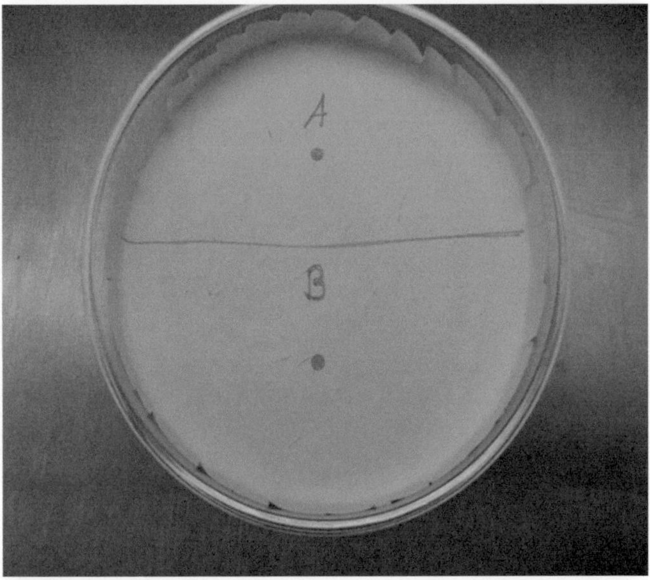

Fig. 5.141 Drying of stained grid

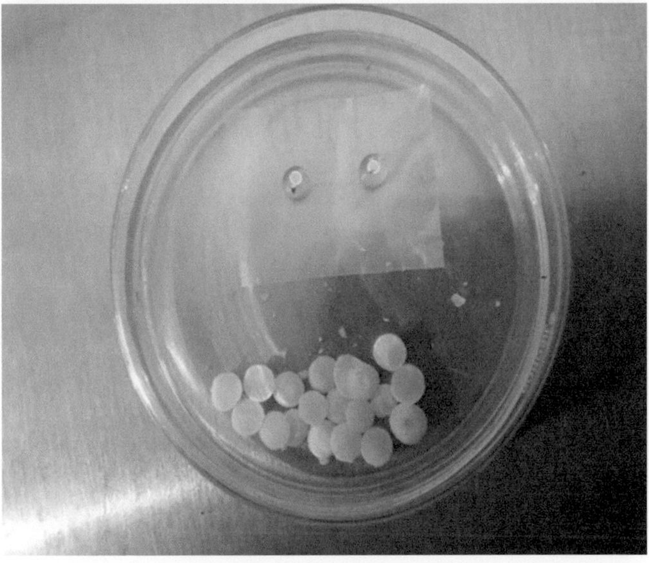

Fig. 5.142 Placing the grid onto the lead citrate stain with section side down and letting the grid float over the stain drop

Fig. 5.143 Covering the petri dish to carry out the staining process in the dark

each stain. This procedure saves much time and has excellent results.

Some TEM micrographs are presented here to understand better the importance of ensuring the proper procedure for cutting and picking up the sections on the grid and the role of staining ultrathin sections before viewing them in the microscope.

A. Error while picking up the sections onto the grid (Figs. 5.146 and 5.147).
B. Good staining (Figs. 5.148 and 5.149).
C. Staining error (Figs. 5.150 and 5.151).

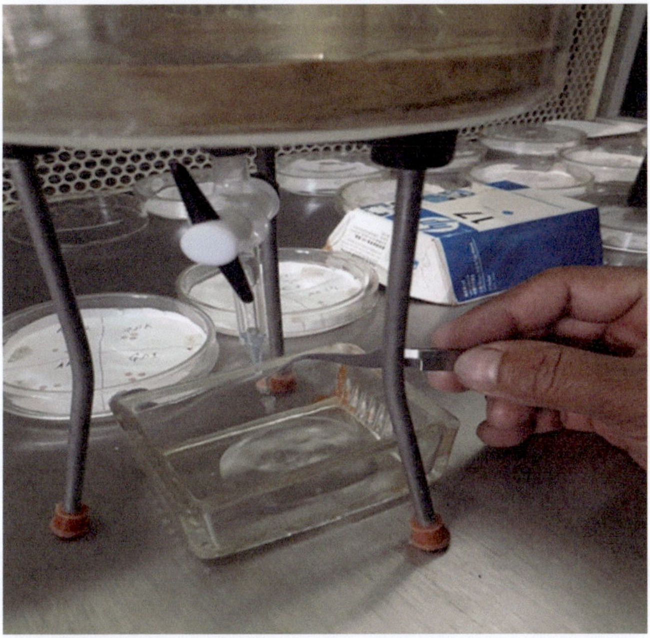

Fig. 5.144 Washing the grid

Storage of Grid-Containing Sections

The grid-containing sections can be stored in a grid box with a grid storage record of different samples or a well-covered Petri dish. Grid boxes protect the grids or sections from outer dust or other artefacts. The sections are now ready for observation in the TEM. Grids can be preserved in a grid box for a long time (Figs. 5.152, 5.153, and 5.154).

Negative Staining

In 1954, Farrant introduced a different method of staining to visualise the fine surface details of small particulate specimens like bacteria, flagella etc., macromolecules like milk proteins, enzymes, haemoglobin, etc., internal cellular organelles like Golgi bodies, nuclear pores, mitochondrial membranes, microtu-

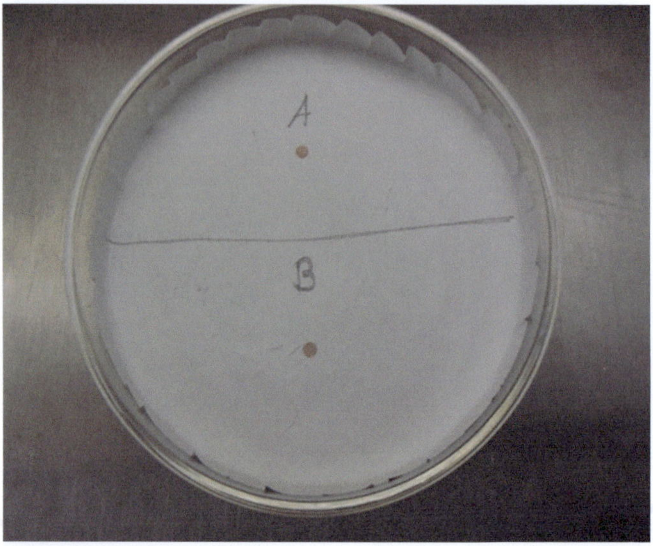

Fig. 5.145 Drying the stained grid

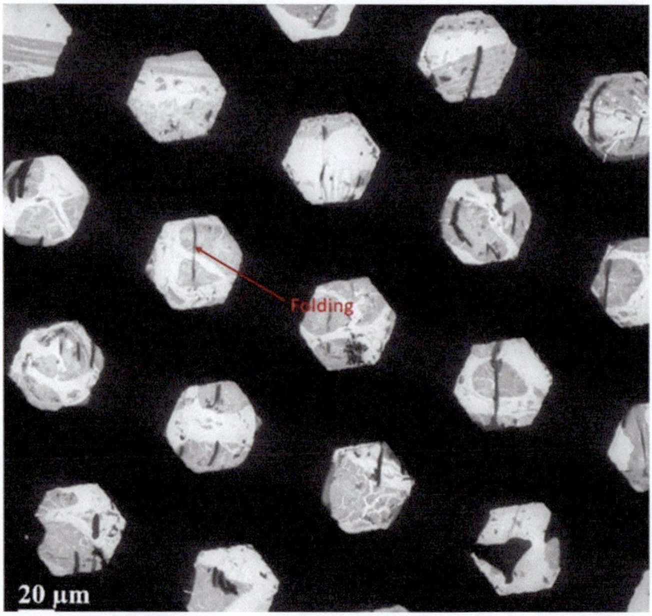

Fig. 5.146 Low magnification micrograph showing folding due to an error in picking up

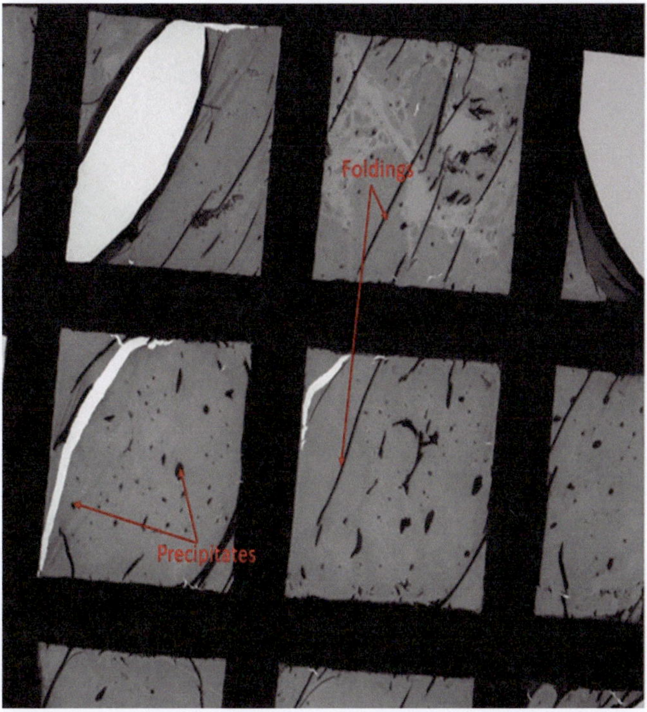

Fig. 5.147 (i) Low magnification micrograph showing foldings due to error in picking up. (ii) The micrograph also shows precipitate formation due to staining error

bules, microfilaments etc. and viruses. This method is called "negative staining" and is an important technique. The method consists of embedding suitably small particulate specimens in a layer of dense material so that the specimen is viewed as a light object against a dark background.

Preparation of particulate specimens for negative staining: 1.0–3.0% solutions of phosphotungstic acid (PTA), ammonium molybdate, uranyl acetate, or methylamine tungstate are used as a negative stain. Solutions of these heavy metals at different pH levels have been effectively used to stain biological specimens. Among these, PTA is the most commonly used stain. To stain with PTA, prepare 2% PTA in doubled distilled water or 0.1–1.0%

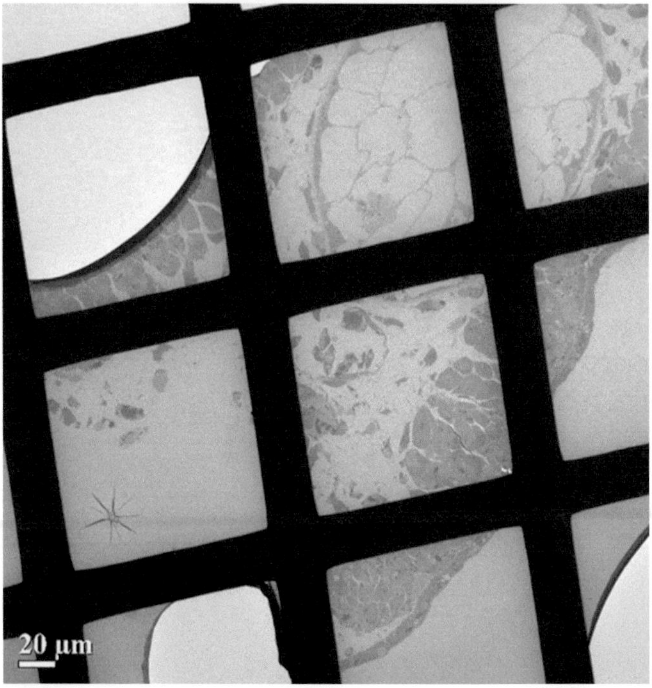

Fig. 5.148 Low magnification micrograph, stained with uranyl acetate followed by lead citrate, depicting good staining

ammonium acetate and adjust the pH between 6.8 and 7.4 with 1 N potassium hydroxide (KOH). The result is potassium phosphotungate, which forms an electron-dense background when dried.

Method of Staining

- Place a drop of the suspension of the particulate specimen (in higher concentration) on a coated grid and drain off the excess suspension with the help of tissue paper or filter paper. At this point, do not allow the suspension to dry on the grid.

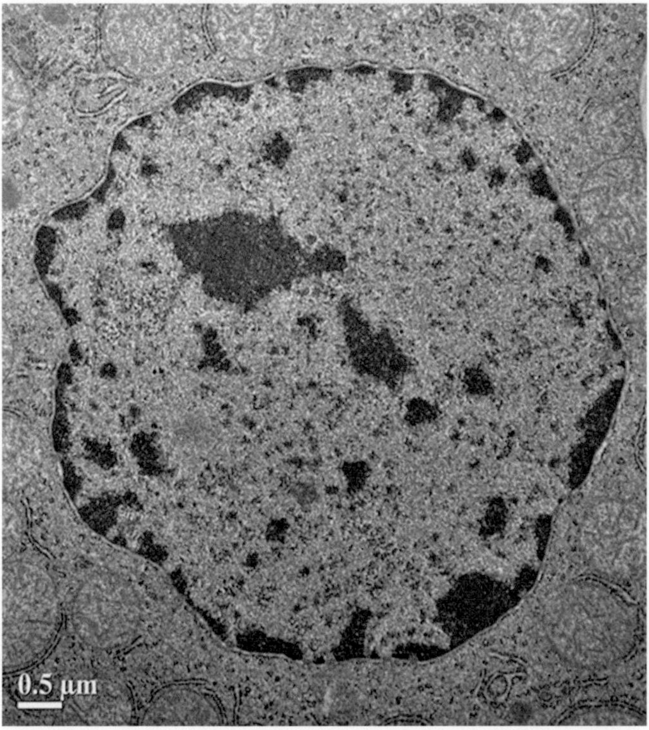

Fig. 5.149 High magnification micrograph, stained with uranyl acetate followed by lead citrate, depicting good staining where the cell organelles are clearly visible

- Continuing, add a drop of PTA or uranyl acetate to the grid and allow to stand for 2 min.
- Now, dry the grid, and the sample is ready to observe under TEM (Figs. 5.155, 5.156, 5.157, 5.158, 5.159, and 5.160).

TEM Viewing

After going through all the steps of block preparation, UT sectioning, and staining, the specimen is now ready to view under TEM

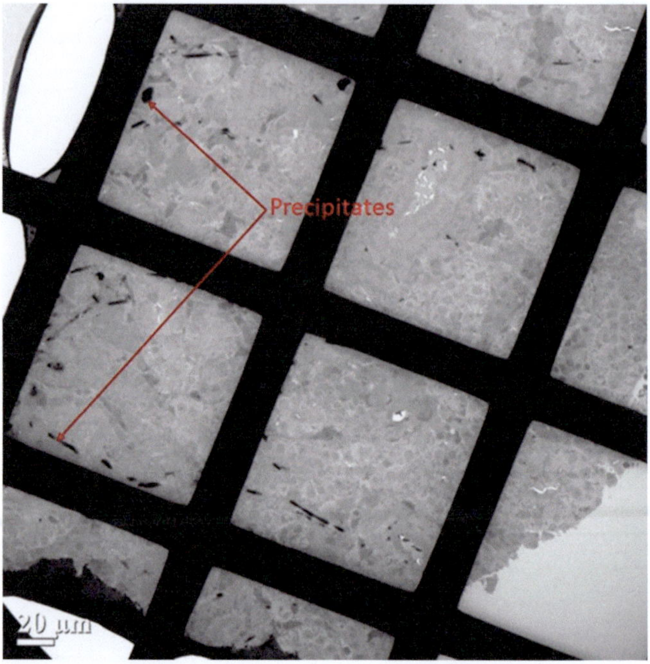

Fig. 5.150 Low magnification micrograph showing precipitate formation due to error in staining

for analysis of its ultrastructure. As mentioned earlier in Chap. 1, TEM can magnify an image a thousand times more than a light microscope with more excellent resolution, which allows for studying the fine cellular details of the specimen. TEM facilitates capturing images in different magnifications, from low to high. Some TEM micrographs from low to high magnifications are given below.

1. Images of UT sections of biological sample (Figs. 5.161, 5.162, 5.163, 5.164, 5.165, and 5.166):
2. Images of UT sections of material sample (Figs. 5.167 and 5.168):

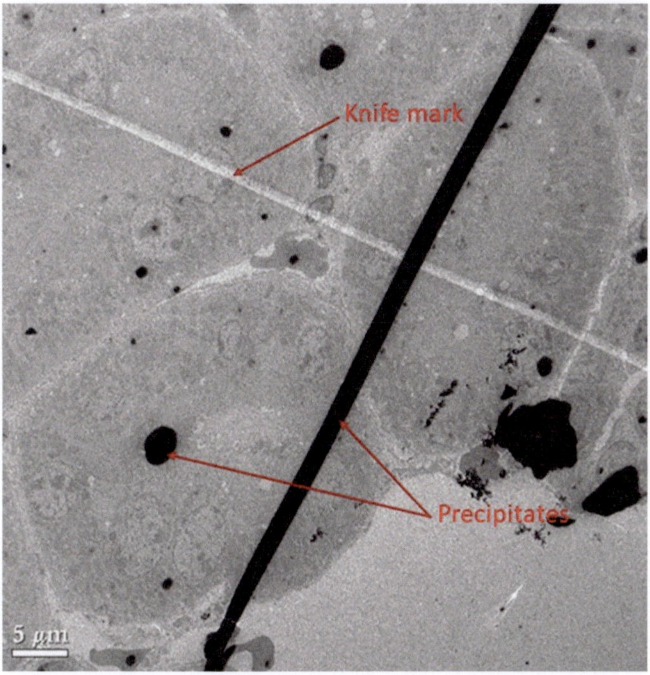

Fig. 5.151 (i) High magnification micrograph showing knife mark due to cutting error. (ii) The micrograph also shows the precipitate formation due to staining error

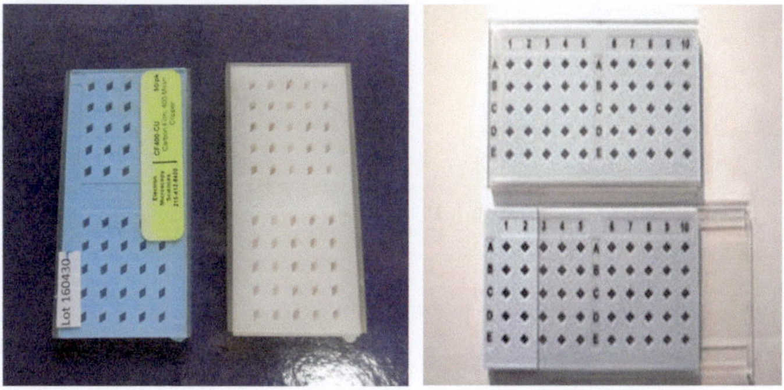

Fig. 5.152 Grid boxes

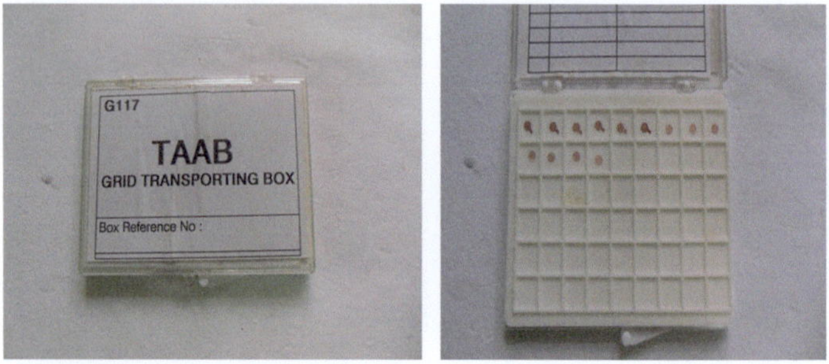

Fig. 5.153 Grid transporting box

GRID RECORD FOR STORING, TRANSPORTING AND VIEWING							
GRID BOX NO.	GRID NO.	LAB CODE	USER CODE	SAMPLE INFORMATION	USER INFORMATION	DATE OF STORING/ TRANSPORTING/ VIEWING	REMARKS

Fig. 5.154 Grid storage record chart

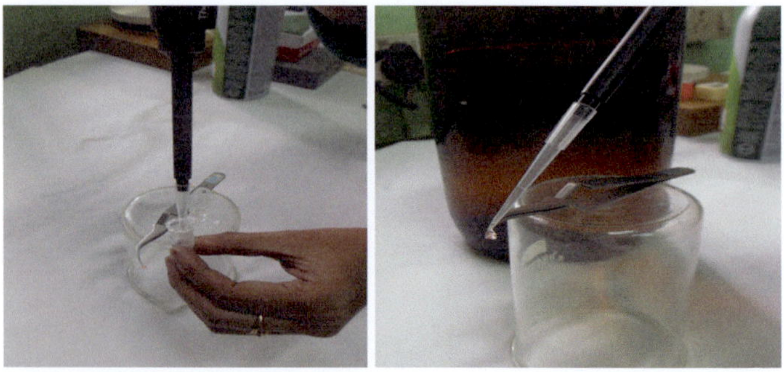

Fig. 5.155 Placing a drop of suspension on the grid

Fig. 5.156 Suspension sample on the grid

Key Learnings

- Specimen block is prepared through the steps, primary fixation → washing with buffer → post-fixation → washing with buffer → dehydration → clearing → infiltration and embedding, to achieve the polymerised block. The blocks are then hand-trimmed by single-edged blades or clean razor blades with a block trimmer to remove the excess resin part surrounding the specimen.
- Glass knives are prepared by special process glass strips and with a glass knife maker, and they are used for machine trimming of the polymerised block, ST sectioning, and UT sectioning.

Fig. 5.157 Draining off the excess suspension

- The quality of the glass knives can be evaluated on a knife evaluation system.
- For cutting ST and UT sections, a trough has to be fixed to the glass knife edge to produce a boat where the sections can be collected later.
- Machine trimming of the specimen block is followed up to smoothen the surface of the block face. The purpose of machine trimming is to make a trapezoid, square or rectangle shape of

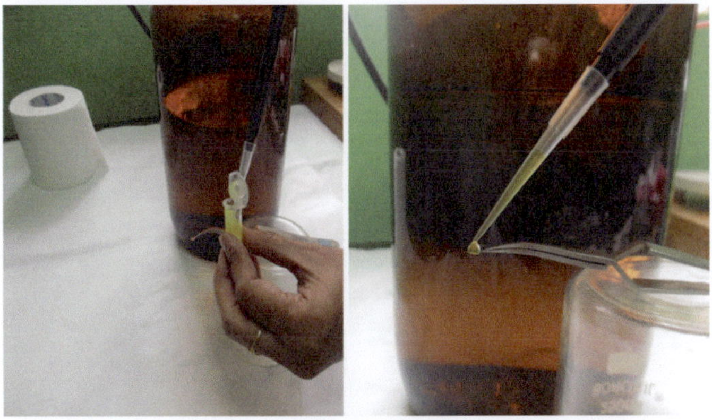

Fig. 5.158 Adding a drop of uranyl acetate to the grid

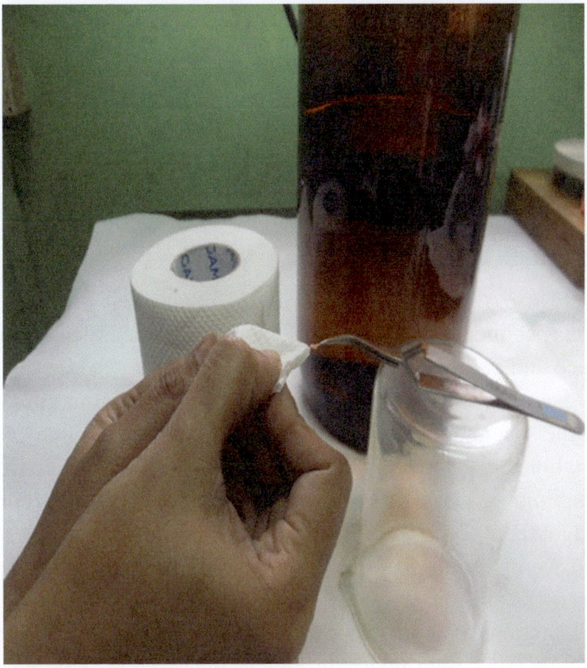

Fig. 5.159 Removing the excess stain

Fig. 5.160 Drying the grid

the specimen on the face of the block. The specimen's smooth face and parallel edges will give serial sections while sectioning, and one section will adhere to another to form a ribbon.

- If the specimen is too big or the sample area is too large, taking a whole area for sectioning is impossible. Moreover, it requires making the sample face small to make it fit the knife's cutting edge; then, some specific trimming tricks have to be followed. So we can minimise the risk of cutting off the required area from the specific sample.
- ST sections of about 400–500 nm thickness are cut for examining the specimen under an optical microscope. This enables determining the status of the block, like the state of the embed-

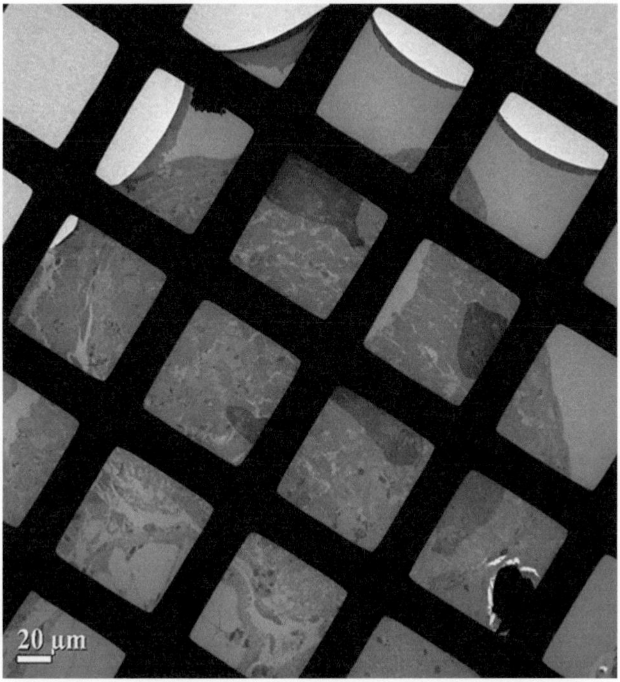

Fig. 5.161 Low magnification TEM image of an ultrathin section showing tadpole head

ding medium (whether soft, medium, or hard) and whether the fixation, dehydration and infiltration of resins are proper (if any holes are present in the tissue). Semithin sections enable us to select the area for final trimming for UT sectioning.

- Staining ST sections will provide enough contrast for better visualisation and distinguish the cellular organelles under an LM.
- Final trimming of the block face before UT sectioning is mandatory for the reason that if the face is too large, like in the case of ST, while cutting the UT sections, the sections may crumple and also, the knife can induce periodic thickness vibrations in

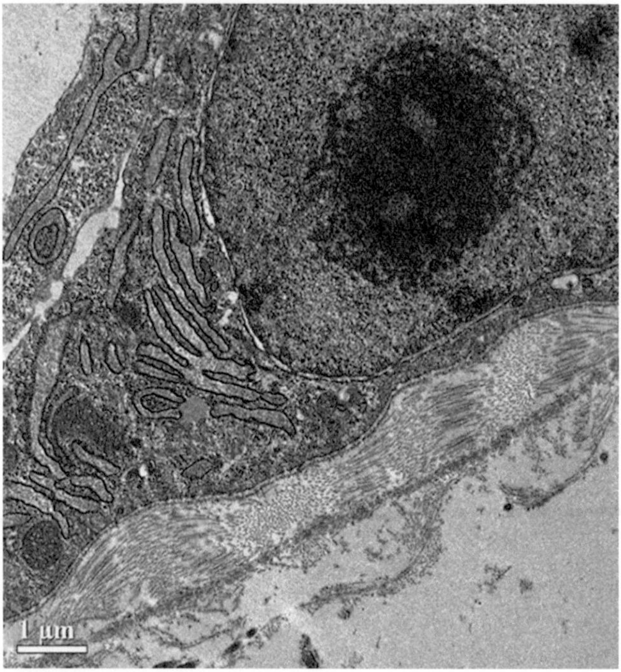

Fig. 5.162 High magnification TEM image of an ultrathin section showing cell membrane, nucleus and ER

the cut that can damage the knife blade itself as the UT sections are very thin (60–90 nm) in nature.

- The diamond knife is one of the most durable yet fragile tools. The diamond edge is exceptionally durable considering the wide range of materials cut successfully from soft to hard. With proper care and handling, the diamond knife can withstand much use.
- The UT sections are cut serially, where one section adheres to the next section, forming a ribbon of sections that floats onto the surface of the water in the boat of the knife. The light reflecting off from the water's surface through the sections gives sections a colour, and so generates colour interference. The interference of colour corresponds to the thickness of the

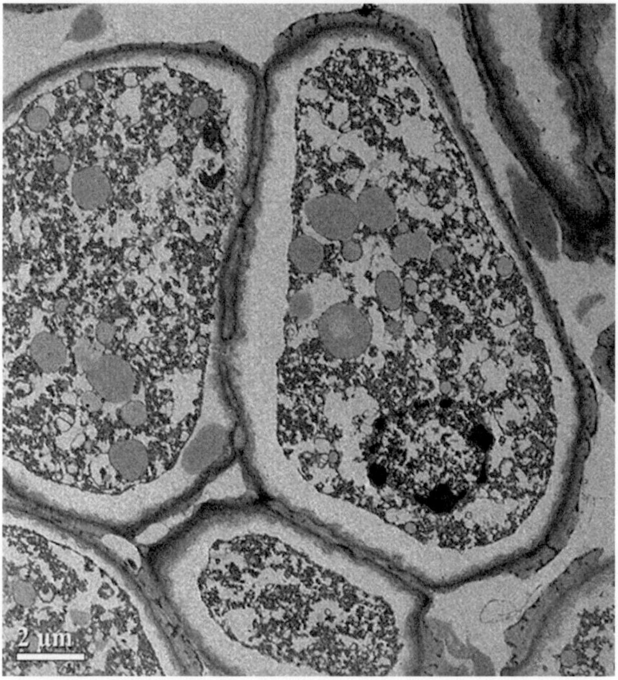

Fig. 5.163 High magnification TEM image showing nucleus, cell wall, and lipids of stressed plant stem

section, enabling the microtomist to know the rough thickness of the section. The colour of the sections floating on the water surface ranges from grey (very thin, below 60 nm) to yellow (very thick, above 280–320 nm). The silver-coloured sections (60–90 nm) have the appropriate thickness for imaging in TEM, and the thicker sections can be used for light microscopy work.

- A good UT section will give an excellent image of the specimen. A small mistake affects the image quality. Some common errors should be avoided during cutting procedures, i.e., knife marks, thick sections, chatter or compression, etc.
- It is necessary to wash and clean up the edge of the knife and knife boat every time after every sectioning procedure is

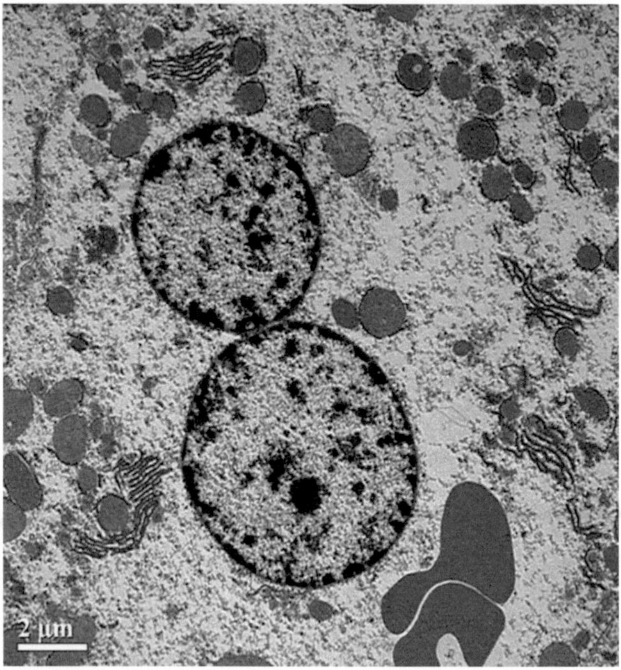

Fig. 5.164 High magnification TEM image of treated liver cell showing abnormal nuclei, ER, mitochondria, and RBC

completed. Different methods are followed to clean up a diamond knife.

- The UT sections are collected on specially made grids for TEM use. Grids are usually coated on one side with specific metals like copper, rhodium, nickel, gold, etc., and are matted on the coated side but bright on the other.
- Different film types are prepared to support the grid. Coated grids also helped to minimise astigmatism in the objective lenses and apertures and manage image stability in TEM.
- The purpose of staining UT sections is to provide contrast to the biological organelles present in the section while viewing in TEM.

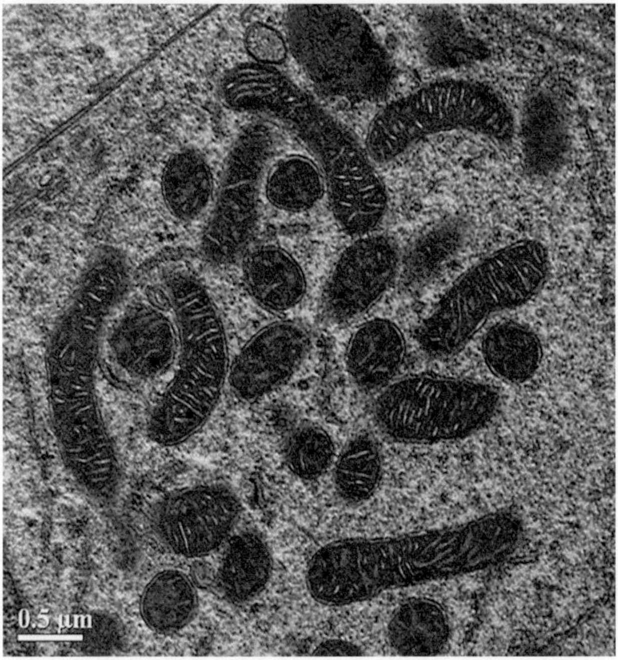

Fig. 5.165 High magnification TEM image of mitochondria showing prominent mitochondrial membranes and cisternae

- The prepared grid can be stored in a grid box, with a grid storage record of different samples or in a well-covered Petri dish.
- Negative staining is used to visualise the fine surface details of small particulate specimens like bacteria, flagella, etc., macromolecules like milk proteins, enzymes, haemoglobin, etc., internal cellular organelles like Golgi bodies, nuclear pores, mitochondrial membranes, microtubules, microfilaments, etc. and viruses.
- After going through all the steps of block preparation, UT sectioning, and staining, the specimen is ready to view under TEM for analysis of its ultrastructure.

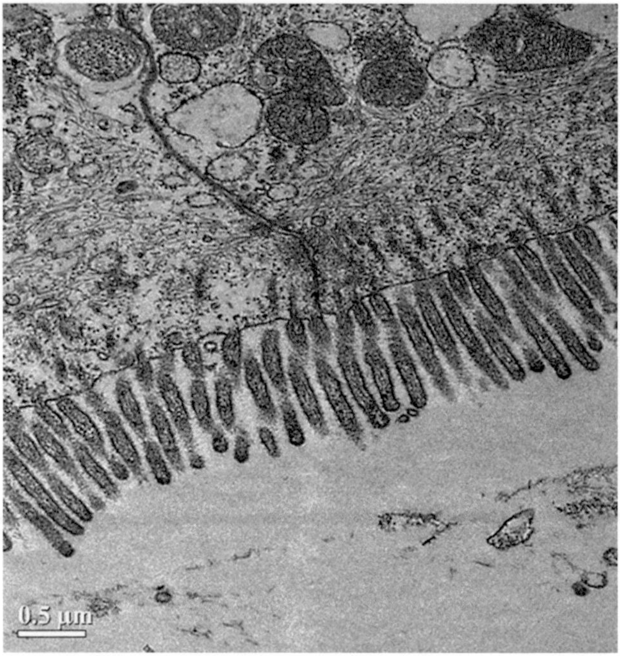

Fig. 5.166 Junction between two epithelial cells of intestine showing microvilli and mitochondria in high magnification

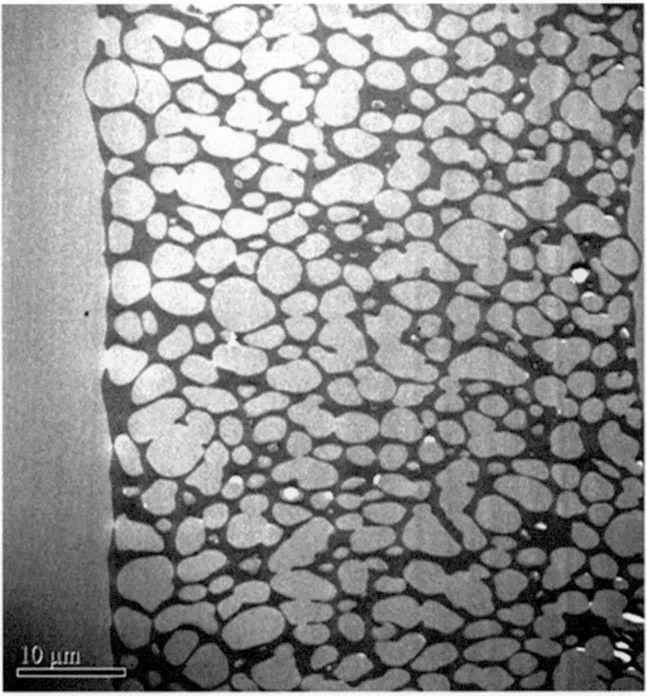

Fig. 5.167 TEM micrograph of polymeric membrane in low magnification

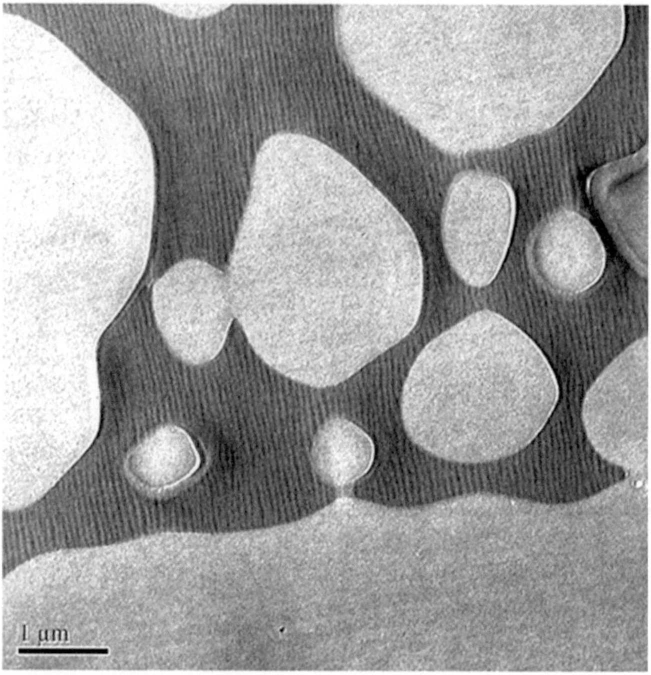

Fig. 5.168 TEM micrograph of polymeric membrane in high magnification

References

Estrada JC, Nathan TB, Bossen EH (1985) A rapid method of staining ultrathin sections for surgical pathology TEM with the use of the microwave oven. Am J Clin Pathol 83(5):639–641

Hall CE, Jakus MA, Schmitt FO (1945) The structure of certain muscle fibrils as revealed by the use of electron stains. J Appl Phys 16:459–465

Hayat MA (1981) Principles and techniques of electron microscopy. Biological applications. Edward Arnold, London

Reynolds ES (1963) The use of lead citrate at high pH as an electron-opaque stain in electron microscopy. J Cell Biol 17(1):208–212

Roland JC, Carole AL, Morre DJ (1972) Phosphotungstic acid-chromic acid as a selective electron-dense stain for plasma membranes of plant cells. Stain Technol 47(4):195–200

6 Common Problems in Ultramicrotome and Its Remedies

With the appropriate cutting method on an ultramicrotome, satisfactory ultrathin sections of about 60–90 nm thickness, corresponding to "silver" colour sections, can be achieved. The regular ultrathin sections are of uniform thickness, without any knife marks or compression and free from impurities.

Learning Objectives from This Chapter

- Irregular sections while cutting
- If the knife edge is not wet or there is difficulty within the knife edge and when the block face gets wet during a microtomy session and also sections dragging
- If the sections are not coming in ribbon form or curving ribbons
- Sections with chatter or compression, vertical lines, and striations
- Knife marks in sections
- If the sections stick to the eyelash brush, sections chipping or fold in section

The frequently confronted problems during sectioning are listed below, along with the causes and remedies;

A. Haloi, *Essentials of Ultramicrotomy*,
https://doi.org/10.1007/978-981-97-4791-7_6

Irregular Sections

- If the sections are not cutting or cutting at an alternate stroke, it may be because of insufficient feed or small section thickness; in that case, increase the feed setting by increasing the section thickness. It may be because the knife edge is not sharp enough, so a freshly prepared new knife should be used. It may also be due to vibration in the surrounding area; then check for it and stop it at the source.
- Suppose the irregularities in thickness or unevenness in the form of bands of different colours appear within one section. In that case, it is usually a symptom of poor infiltration of the specimen, improper resin polymerisation, or may be due to vibrations nearby. In such a situation, block the vibration and check if any locking knob is not properly secured and then secure it properly.
- If each section is cut with a different thickness, usually altering the section thickness from thick to thin in each cut, then all locking controls have to be checked to see whether it is locked properly; if found loose, then tighten it to secure the locking system. If the problem remains, check for any vibration around, then stop it. Sometimes, this problem persists after solving the previous two remedies; in that case, increase the section thickness by a few steps or thicken the thickness for the initial few sections and then gradually reduce the thickness until it appears as a silver colour section.

The Knife Edge Is Not Wet or Difficulties Within the Knife Edge

- While cutting sections, if the knife edge is not getting wet, it may be because of the change in humidity or the diamond knife is not in use for a longer period; in that case, put a few more drops of distilled water, forming a convex surface on the boat then pull the water level to the knife edge with the help of single eyelash brush and leave in this position for few minutes then carefully remove the excess water until it becomes the same level with a knife edge or the reflection of light can be seen on the edge of the knife.

- If the first process fails to wet the edge of the knife, pure alcohol (100%) is used to run over the edge of the knife or wet the knife edge first with alcohol or the knife edge can be cleaned with 100% alcohol with a cleaning rod, then fill the boat with distilled water for sectioning of the specimen.
- Another solution is when the knife is mounted, and the boat is being filled, the knife cleaning rod or a hair or eyelash brush should pass over the cutting edge gently, using the same motion to make the edge wet.

Wet Block Face

- Block face sometimes might get wet due to electrostatic charging, which causes low room humidity and transportation. Another reason is that some of the embedding medium is hydrophilic, so it tends to attract water. In that case, increase the room humidity.
- If the block face becomes wet while cutting sections and while the block passes through the knife edge, the water level on the trough may be on the higher side; in that case, lowering the water level slightly with the help of trough filler or a syringe while knife edge is still wet will alleviate this problem. It may be noted that before cutting again, the block face and the back of the knife's edge must be blotted dry with a piece of filter paper until all the water has been removed properly.
- The embedding material may be too soft. For example, acrylic resin blocks easily pick up the water from the boat. The cutting speed should be increased to solve this problem.
- If the above two remedies fail, then check for the given clearance angle, which may be incorrect, and then try another clearance angle.
- Check for the orientation head, which may be locked at the incorrect height, then correct it by adjusting the knife and the

specimen face locked at the height of the knife height indicator.

- The cutting zone may be set incorrectly for the size of the specimen face, and then it is corrected by setting another cutting zone.

Sections Are Not Coming in Ribbon Form

- If the serial sections are not forming a ribbon, it may be because the specimen edges are not smooth; a retrim of the specimen face is required.
- It may be because the top and base of the specimen face are not parallel; in that case, a retrim of the top and base edges is recommended.
- It can also be because the top and base of the specimen are not parallel to the knife edge; therefore, realignment of the specimen face to the knife edge is needed.

Curving Ribbons

- While cutting sections, if the ribbon of sections moves or curves towards any one side of the boat, the bottom edge of the block face may not be parallel to the knife edge, so the specimen face should slightly rotate to get it correct.
- After correction, if the problem continues, it may be because the top and bottom edges of the block are not parallel. Then, it is advised to remove the block from the holder and retrim it. Also, pick off a few initial cut sections at a time so they do not become curved at the knife's edge.

Sections with Chatter or Compression

- Chatter is compression, generally visible only on an electron microscope. It cannot be seen through the stereomicroscope of the ultramicrotome as the section expands in the boat shortly after the cut, and it consists of very close, evenly spaced bands of thick and thin areas present on the sections. It may be due to the knife's edge dullness, so a knife change is recommended.

- If the chatter still appears, increase the knife angle by 1° and another 1°, if needed.
- Another reason may be that the surface tension of the trough liquid is too high, so reduce the surface tension by adding alcohol to the trough liquid.
- It may be because the cutting parameters are incorrect; in that case, reduce cutting speed, knife clearance angle, and knife angle.
- If changing the parameters does not work out, poor embedding may be a reason; try to stretch the sections with chloroform vapour.
- It may also occur due to external vibrations.
- The screws of the block, block holder, and knife are not fully tightened. In that case, make sure that all the screws are correctly tightened.
- Cutting pressure is too big. This may be solved by reducing the block width.
- It may happen due to a faulty microtome. In that case, the microtome can be checked by a service engineer.
- It may also be because of,

A. If the block is too soft—make the block harder by reheating it.
B. If the knife angle is too big—Switch the knife angle from 45° to 35°.
C. If the knife is dull—If possible, resharpen it or change the knife.
D. The clearance angle is too big, and the cutting speed is too high—reduce the clearance angle by 1–2° and reduce the cutting speed gradually.

Vertical Lines

- If vertical lines or scratches are seen on the section, the knife edge may be damaged, and a fresh glass knife should be used.

For a diamond knife, change the cutting area to another position or carefully clean the knife's edge.
- It may be because of the presence of hard particles on the specimen face, then trim off the hard particles and make the specimen face smaller.

Knife Marks

- Knife marks are tears or streaks parallel to the direction of sectioning and generally with uneven spacing and width, causing very fine lines in the sections, mainly caused by sections stuck or dried on the knife's edge and not cleaned properly. The knife marks are the main reason for the resharpening of the diamond knife. These may be because of the dullness of the cutting edge or debris on the knife edge or in the resin; in such cases, the cutting area of the knife edge should be changed, or the knife should be changed.
- It may be due to hard material in the specimen and hard specimens like bone. That is why semithin sections should always be cut before processing for ultrathin sections, and the glass knife should be used for semithin sections rather than a diamond knife for ultrathin sections. Hard materials are frequently present in cell suspensions if proper care is not taken during sample collection, processing, and embedding. So, if the knife marks are seen in semithin sections, that particular area from the face of the block can be trimmed off before cutting the ultrathin sections. If that area is of vital interest, then ultrathin sections can be cut with glass knives, and the micrographs can be taken from around the knife marks.
- The reason may be the presence of debris in embedding media; in this case, re-embedding the specimen in fresh embedding media is recommended.
- Touching the knife edge with any solid object or by mistake with fingers can cause knife spotting, and later, while cutting sections, in that spot, the sections might split or cause small or big knife marks according to the proportion of damage on the

knife. So, to solve this, all hard objects and fingers should be kept away from the knife edge.

- Remnant particles remain in the block while trimming—the block face should be appropriately brushed to remove all the trimmed sections from its face before cutting.
- Any hard particles or artefacts present in the block while processing due to human error—keep the highest possible purity during the entire sample processing or block preparation procedure from primary fixation to the specimen embedding. There are high chances of getting artefacts in the case of suspension samples (blood, sperm, microorganisms, algae, etc.) since the sample processing procedure has to undergo centrifuging every time, so it needs to follow an immaculate way.
- Regularly using a diamond knife for a very long period and sectioning different specimens every time will also cause the nick at the edge.

Sections Sticking to Eyelash Brush

- While picking up and moving the sections around, if the sections get stuck to the single eyelash brush as the brush has to remain clean for every use, then hold a piece of clean filter paper while placing the single hair brush in between the paper and then pull it, the stuck sections will be removed.
- Another way to do it is to run the thumb finger and forefinger down to the nose, then pull the single hair brush between these two fingers.

Section Dragging

- While cutting sections, if the sections are dragged out of the knife with every cutting stroke and come towards the back side of the knife, then the block face may slightly catch up with the water from the boat of the knife. Similar remedies are followed in that case, as in the case of a wet block face.

Sections Folded

- There are two processes of picking up the sections: one from above and another from below the water. This problem is mainly seen if the sections are picked up from above the water. So, sections will be picked from below following the procedure mentioned in Chap. 5.
- Sometimes, this occurrence is because of overstretched sections with chloroform vapour. In that case, section stretching for a longer time with chloroform should be avoided. Overstretching of sections also causes ruptures in sections, which is only seen during viewing in TEM.

Contamination on Sections

- If the sections are evidenced with dirt, it may be from trough liquid or distilled water. Then, the water from the syringe or trough filler has to be changed, and fresh trough water is used, and if needed, a new knife is used with the trough. If it is a diamond knife, it has to be washed thoroughly.
- It may be from the adhesive used while fixing the trough, which is dissolved; in this case, use a fresh knife.

Sections Chipping

- If the embedding media of the block is too hard, then the sections are used to chip into the knife's edge. The solution to this problem is to change the cutting parameters by reducing the clearance angle of the knife and cutting speed.

Striations

- Striations are usually found in the sections themself—serration or very tiny imperfections at the knife's edge—Try to cut a section from another area of the knife.
- If the block is not correctly or poorly polymerised, make sure the block is fully polymerised.
- Inhomogeneous block or no proper consistency—Change the block or take a duplicate block of the same specimen.

Maintenance of Ultramicrotome

- The ultramicrotome may be installed on the ground floor in a vibration-free room.
- Regular cleaning and brushing off the microtome cutting area after every microtomy session is very important.
- Apart from using a dehumidifier in humid weather, a table lamp may be illuminated near the microtome for extended microtome longevity.
- The water connection and wash basin may fit far away from the microtome area and, if possible, in a nearby room for the long life of the microtome.

Good ultrathin sections can be achieved by correctly applying all the parameters of ultramicrotomy, which is the primary requisite to view the ultrastructure of cell organelles of biological samples in TEM. However, good ultrathin sections depend on a good block, meaning a block with even consistency. If all the parameters are correctly done in all the steps, starting from primary fixation, secondary fixation, dehydration, clearing, polymerisation, and embedding with minimised human error or no error, then a good embedded block with even consistency can be found. Biological sample preparation for TEM is a lengthy, crucial, tedious, and laborious process, and the quality of the TEM micrograph is highly dependent on it.

Key Learnings

- Irregular sections may be resolved by checking a few parameters, for instance, by increasing the section thickness or using a freshly prepared new knife or securing the locking knob properly or obstructing the nearby vibration.
- If the knife edge does not get wet while cutting sections, pure alcohol may be used to run over the edge of the knife or the boat may fill with a few more drops of water till it touches the edge and wait in this position to stabilise, then remove the excess water. If the block face gets wet during the microtomy session or if the sections are dragged out of the knife boat with every cutting stroke, this problem may be resolved by increas-

ing the humidity of the room, slightly lowering the water level from the trough, increasing the cutting speed, by changing the clearance angle or by setting another cutting zone.

- If the serial sections are not forming a ribbon, realignment of the specimen face to the knife edge may needed or retrim of the specimen face. If the ribbon of sections is curving, the specimen face should slightly rotate to get a correct alignment or pick off a few initial cut sections at a time.
- If chatter appears during sectioning, increase the knife angle by 1°, reheat the block, and reduce the clearance angle by 1–2° or reduce the cutting speed. If the vertical lines are seen on the section, a fresh glass knife should be used, or in the case of a diamond knife, change the cutting area to another position or trim off some area and make the specimen face smaller in size. If serration or very tiny imperfections at the knife's edge occur, then sections may be cut from another area of the knife or the block.
- If knife marks appear, the cutting area of the knife edge should be changed, or the knife should be changed. The block face should be appropriately brushed to remove all the trimmed sections from its face before cutting.
- If the sections get stuck to the eyelash brush, they can be removed by holding a piece of clean filter paper, placing the brush between the paper, and then pulling it.
- Folding on the sections is seen mostly if the sections are picked up from above the water. So, sections are to be picked from below the water following the proper procedure. Also, overstretching of sections can cause ruptures in sections, which is only seen during viewing in TEM.
- If the sections chip to the knife's edge, then change the cutting parameters by reducing the clearance angle of the knife and cutting speed and also wipe off the block face.
- Regular maintenance of the ultramicrotome is essential for its long life.

Correction to: Microtome Device and Microtomy

Correction to:
Chapter 3 in: A. Haloi, *Essentials of Ultramicrotomy*,
https://doi.org/10.1007/978-981-97-4791-7_3

The original version of this book was inadvertently published with an incorrect figure in Chapter 3. The following corrections have been made in the book after publication.

In Chapter 3 (p. 71), the figure has been corrected as follows:

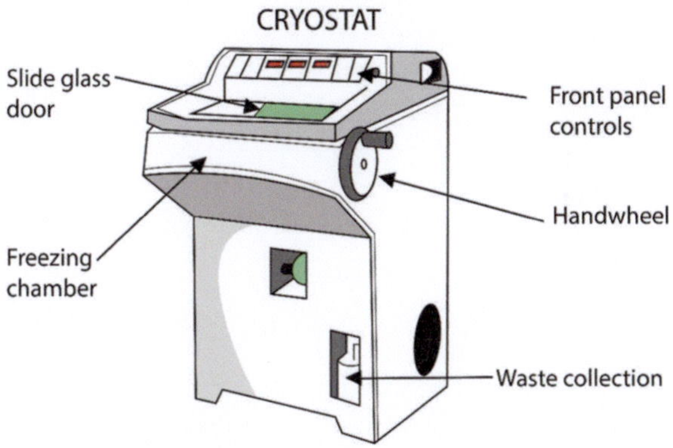

The corrections have been carried out in the chapter.

The updated version of this chapter can be found at
https://doi.org/10.1007/978-981-97-4791-7_3

A. Haloi, *Essentials of Ultramicrotomy*,
https://doi.org/10.1007/978-981-97-4791-7_7

Appendices

Appendix A

Taking appropriate laboratory safety precautions is necessary in appropriate areas of sample preparation, which has been addressed in the text of this book. In an electron microscopy laboratory, sources of hazards and injuries are exposure to various chemicals, including some volatile and heavy metals, hot objects like pumps, multiplates, hot plates, incubators, etc., glass and razor blades, and inflammable materials, etc. If the chemicals are properly handled, the chance of personal injury is reduced as they are used in small proportions in sample preparation for electron microscopy, even though they are toxic. The sink for washing and the fume hood should be readily available to handle the carcinogens, and eye wash stations and showers for spills should be available within the laboratory. Also, hands and eyes need to be protected at all times. Additionally, the general laboratory safety guidelines are to be followed.

Appendix B

Various books and journals have been referred throughout the text with the aim of providing relevant information and starting research questions about the technique followed for microscopy and ultramicrotomy procedures in TEM.

A. Haloi, *Essentials of Ultramicrotomy*,
https://doi.org/10.1007/978-981-97-4791-7

Books

Avers CJ (1978) Basic cell biology. Van Nostrand Reinhold Inc, New York

Bozzola JJ, Russel LD (1999) Electron microscopy: principles and techniques for biologists. Jones and Barlett Publishers, Sudbury, pp 48–71

Dykstra MJ (1992) Biological electron microscopy, theory, techniques, and troubleshooting. Plenum Press, New York

Hayat MA (1981) Principles and techniques of electron microscopy. Biological applications. Edward Arnold, London

Hayat MA (1984) Fixation for electron microscopy. Academic Press, Inc, New York, pp 320–331

Hill J (1770) The construction of timber, from its early growth, explained by microscope and proven from experiments in a great variety of kinds. The author, London, pp 5–11. Plate I

Verma PS, Agarwal VK (1999) Cell biology, genetics, molecular biology, evolution and ecology. S. Chand & Company LTD, Ram Nagar, New Delhi

Journals

Dey S, TSB B, Roy B, Dey D (1989) A new rapid method of air-drying for scanning electron microscopy using tetramethylsilane. J Microsc 156(2):259–261

Estrada JC, Nathan TB, Bossen EH (1985) A rapid method of staining ultrathin sections for surgical pathology TEM with the use of the microwave oven. Am J Clin Pathol 83(5):639–641

Farrant JL (1954) An electron microscopic study of ferritin. Biochim Biophys Acta 13(4):569–576

Fernandez-Moran H (1953) A diamond knife for ultrathin sectioning. Exp Cell Res 5(1):255–256

Hall CE, Jakus MA, Schmitt FO (1945) The structure of certain muscle fibrils as revealed by the use of electron stains. J Appl Phys 16:459–465

Hohl HL, Hamamoto ST, Hemmes DE (1968) Ultrastructural aspects of cell elongation, cellulose synthesis, and spore differentiation in *Acytostelium leptosomum*, a cellular slime mould. Amer J Bot 55(7):783–796

Karnovsky MJ (1965) A formaldehyde-glutaraldehyde fixative of high osmolarity for use in electron microscopy. J Cell Biol 27:137–138

Latta H, Hartmann JF (1950) Use of a glass edge in thin sectioning for electron microscopy. Proc Soc Exp Biol Med 74(2):436–439

Reynolds ES (1963) The use of lead citrate at high pH as an electron-opaque stain in electron microscopy. J Cell Biol 17(1):208–212

Roland JC, Carole AL, Morre DJ (1972) Phosphotungstic acid-chromic acid as a selective electron-dense stain for plasma membranes of plant cells. Stain Technol 47(4):195–200

Stein RS, Norris FH (1956) The X-ray diffraction, birefringence, and infrared dichroism of stretched polyethylene. Pol Sci 21(99):381–396

The Nobel Prize in Physics 1986 Perspectives—Life through a lens. nobelprize.org

MIX
Papier aus verantwortungsvollen Quellen
Paper from responsible sources
FSC® C105338

If you have any concerns about our products,
you can contact us on
ProductSafety@springernature.com

In case Publisher is established outside the EU,
the EU authorized representative is:
Springer Nature Customer Service Center GmbH
Europaplatz 3, 69115 Heidelberg, Germany

Printed by Libri Plureos GmbH
in Hamburg, Germany